Computing and Systems Technology Division 2014

Core Programming Area at the 2014 AIChE Spring Meeting and 10th Global Congress on Process Safety

New Orleans, Louisiana, USA
30 March – 3 April 2014

ISBN: 978-1-63439-068-2

Printed from e-media with permission by:

Curran Associates, Inc.
57 Morehouse Lane
Red Hook, NY 12571

Some format issues inherent in the e-media version may also appear in this print version.

Printed by Curran Associates, Inc. (2014)

AIChE
3 Park Avenue
New York, NY 10016-5991

Phone: (203) 702-7660
Fax: (203) 775-5177

www.aiche.org

Additional copies of this publication are available from:

Curran Associates, Inc.
57 Morehouse Lane
Red Hook, NY 12571 USA
Phone: 845-758-0400
Fax: 845-758-2634
Email: curran@proceedings.com
Web: www.proceedings.com

TABLE OF CONTENTS

(48a) Change Your Board Operator to a Process Manager with State-Based Control 1
Tom Nolan and Dustin Beebe
(48aa) Huaa – When Learning Is Not Enough 16
Mike Bearrow
(48ac) Autocad - Smart Grid – Agriculture - Architecture and the United States Government 20
Ethenia Scott
(48ad) Production, Characterization and Catalytic Studies of Biobased Carbon Materials 21
Qiangu Yan, Jilei Zhang and Zhiyong Cai
(48ag) Uncertainty in Sour Gas Viscosities Estimation, What Is the Effect on Your Reservoir Inflow and Tubing Performance 22
Adel Elsharkawy
(48ah) Chromonic Nanocarriers for Chemotherapeutics: Size Distribution and Control Release Studies 23
Rahul Misra and Sanat Mohanty
(48ai) Double Containment Piping Solutions for Safety and Environmental Concerns 24
Patrick Fedor and Darin Johnson
(48b) DTP R Process: On-purpose Propylene Production Technology 31
Kazunori Honda, Atsushi Okita, Jumpei Takahashi, Koji Oyama, Nobuyasu Chikamatsu, Mitsuo Morita and Shuji Obayashi
(48d) Self-Cleaning "Bernoulli" Type Filters Used in Onshore and Offshore Applications 36
Artur W. Krueger
(48f) Effects of Support on Sulfur Tolerance and Regeneration of Pt Catalysts Measured By Ethylene Hydrogenation and EXAFS 37
Jorge Pazmino, Chuansheng Bai, Jeffrey T. Miller, Fabio H. Ribeiro and W.N. Delgass
(48h) Improved Operational Efficiency and Reliability through Insulation Materials Selection 38
Steven Coppella
(48ij) Compliance with EPA Boiler MACT standards: Mercury-In-Fuel Gas 39
Patrick Laine
(48k) Young Professional Simulation Tutorial 40
Naomi Hua and Mike Donahue
(48m) Development Of Polymeric Sulfonic Acid Composite Membranes For Fuel Cell Applications 44
Jimoh Adewole, Abdullah S. Sultan, Amir Al-Ahmed and S. M. Javid Zaidi
(48o) Facility Siting for Major Projects – Implementation of Consequence Analysis/Quantitative Risk Analysis, a Project Development Lifecycle Framework 52
Mohammad Faruq Haider
(48p) Global Energy and Transportation and Mobil Oil 65
Ethenia Scott
(48r) Characterization of Iron Phthalocyanine As the Active Material for Lithium Batteries 66
Sarwan S. Sandhu, Joseph P. Fellner and David Anneken
(48s) Design of a Free-Fall Reactor for Fast Pyrolysis of Waste Plastics 67
Pravin Kannan and Ahmed AlShoaibi
(48t) Biosorptive Dehydration of Ethanol/Water Azeotropes Using Compound Starch-Based Adsorbent 68
Wenping Wang, Jinsheng Sun, Xijia Cao and Guangxin Liu
(48u) Purification of 2-Amino-1-Phenylethanol Enantiomers By a New Technique Combining Distillation and Crystallization 70
Lie-Ding Shiau and Hou-Guo Teng
(48w) Fabrication of Low Cost Insulating Material from Kaolin Clay for Construction Purposes 71
Naim Faqir, MA Al-Harthi, Hamad AbdulWahhab, Mazen Alshaaer and Reyad Shawabkeh
(48x) Thermodynamic Study of Binary PAH (Anthracene + Phenanthrene) Solid Mixtures 72
James W. Rice, Jinxia Fu, Emma Sandström and Eric M. Suuberg
(48y) How Confident Are You That Your Major Accident Risks Are Under Control? 73
Ellis Graeme and Robert Smith
(48z) Sil Determination: Shortcomings with the Use of LOPA 74
Alan King
(117a) Chemical Process Development - a Lean Six Sigma Perspective 76
Darren Clark

(117b) Using Six Sigma and Experimental Design in Process Development ... 77
Mark V.M. Emonds
(117c) Lean Six Sigma Methods in Xerox Corporation ... 90
Tom Enright
(117d) Development of Continuous Improvement Network in Xerox Manufacturing ... 100
Terry Missana and Tom Enright
(117e) Use of Six Sigma Tools in the Commercial Scale-up of a Novel PPE Copolymer ... 101
Alvaro Carrillo
(136a) A Qualitative in Field Test to Characterize Oil-Shale Reservoirs ... 102
Roy A Drayton
(136b) Pressure Regulation and Kick Attenuation with Wired Pipe Technology in Managed Pressure Drilling ... 103
Hector Perez, Reza Asgharzadeh Shishavan, James Memmott, John D. Hedengren and David Pixton
(136c) Risks from Unconventional Wells Gasification ... 115
Dennis Butts
(136d) Supercritical CO_2 / Brine / Hydrocarbon Interactions in Shale Rock Fracture: A Geo-Microfluidics Approach ... 124
Peng He, Mark L. Porter, J. W. Carey and Hari Viswanathan
(136e) Model Shale Gas Properties from a Statistical Mechanical Perturbed Hard-Sphere Chain Cubic Equation of State ... 125
Arthur S. Gow, Anna O,Malley, Tyler Trimble, Robert Kelly and Saad Al Khaldi
Author Index

PROSYS™ Maximize Your Plant.

Operator Errors and What Can be Done to Minimize

Global Congress on Process Safety
March 31, 2014

10TH GLOBAL CONGRESS ON PROCESS SAFETY

PROSYS Maximize Your Plant.™

Presenter

Tom Nolan

- Graduated from Ohio University with B.S. in Chemical Engineering
- 24 years experience in chemical process industry in a variety of roles
- MAIC and DFSS Six Sigma Black Belt

PROSYS Maximize Your Plant™

The Cost of Errors

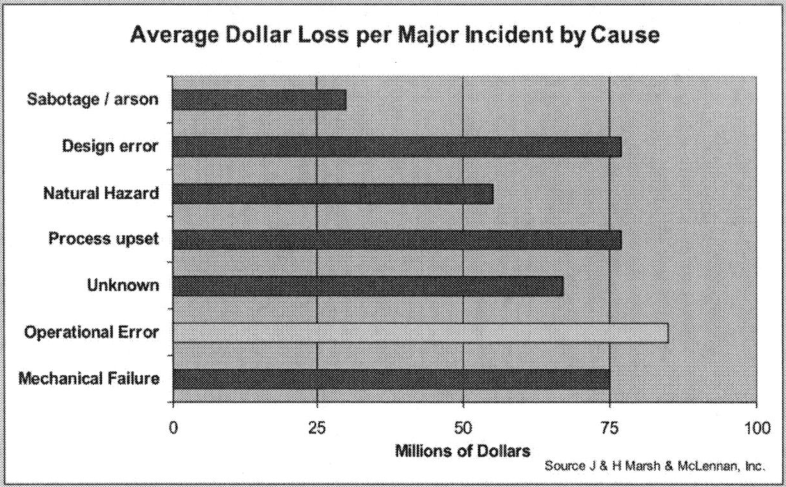

PROSYS Maximize Your Plant™

The Cost of Operator Errors

- ASM estimates total loss due to operator error is $8B per year
- Chemical Safety Topical Committee - average of one chemical incident per day - cost of over $2 million per incident to <u>comply with requirements</u>
- Errors cause 42% of unscheduled shutdowns
- 70% of process incidents occur during start-up or shutdown

PROSYS Maximize Your Plant.™

Reasons for Errors

- Lack of Skill
- Lack of Knowledge
- Carelessness
- System Design
  - Operator set up to make errors by inappropriate design – built in errors
  - Errors that are predictable are preventable by better design

PROSYS Maximize Your Plant.™

Focus of this Discussion

- System Design
  - Alarm Management
  - Operator Graphics
  - Operator set up to make errors by inappropriate design – built in errors
  - Errors that are predictable are preventable by better design

PROSYS Maximize Your Plant.™

10th Global Congress on Process Safety

Alarm Floods

In a number of industrial incidents, alarm floods were identified as a significant contributing cause to the incident…

As found by EEMUA in 1999 and CSB

Alarm Flood defined by ISA 18.2 as -
"10 or more annunciated alarms in any 10 minute period per operator"

PROSYS Maximize Your Plant.™

10th Global Congress on Process Safety

Why do Alarm Floods Occur?

One reason is not providing dynamic alarm management

- **Alarms need to indicate abnormal situations that require operator action**
- **Processes do not operate in one state**
- **What is normal vs abnormal changes with state**
- **Alarms are typically configured for run therefore many alarms are triggered upon a Change of State: Run to Shutdown**
- **Many of the alarms are not applicable or actionable for the new state impleading the operators ability to act quickly on what is important**

PROSYS Maximize Your Plant.™

What is Impacted by Alarm Floods?

- Product quality
- Operability or profitability of the process
- Loss of equipment
- Loss of containment – environmental releases
- Injury and loss of life in plant or community

PROSYS Maximize Your Plant.™

What Makes Alarm Floods so Dangerous?

Can be a problem for three reasons:

- A deluge of alarms can cause critical alarms to be missed
- Floods can be a significant distraction when dealing with process upsets
- Can be an indicator of larger systemic safety issues

PROSYS Maximize Your Plant.™

10th Global Congress on Process Safety

Impacting Alarm Management Design

- **Alarm rationalization is not a process to eliminate alarms - its about quality**
- **Good rationalization will add alarms when appropriate**
- **Typical (Static) rationalization is only for run mode**
- **Dynamic rationalization considers all plant modes**

PROSYS Maximize Your Plant.™

10th Global Congress on Process Safety

Dynamic Alarm Management

- **Dynamic rationalization does everything a static rationalization does plus asks "when" for each alarm**
- **Requires dynamic software to make changes based upon operating mode of the plant**
- **Eliminates redundant alarms and lowers operator loading during transitions**
- **Only alarms what is abnormal and actionable for the given state**

PROSYS Maximize Your Plant.™

10th GLOBAL CONGRESS ON PROCESS SAFETY

Actual Performance Metrics vs ISA 18.2

Type	Avg Alarm Rate/hr	Low/Hi of Avg Rate/hr	Peak Alarm Rate/hr	Low/Hi of Peak Rate/hr	% Time in Flood	Standing Alarms >24 hrs
Before Rationalization	30	Low=6.2 Hi=61	638	Low=152 Hi=2402	17.2%	9
ISA 18.2 Target Metrics	6	---	<60	---	<1%	<5
After Dynamic Alarm Management	2	Low=0.09 Hi=2.5	25	Low=13 Hi=42	0.25%	4

PROSYS Maximize Your Plant.™

10th GLOBAL CONGRESS ON PROCESS SAFETY

Results After Dynamic Alarm Management

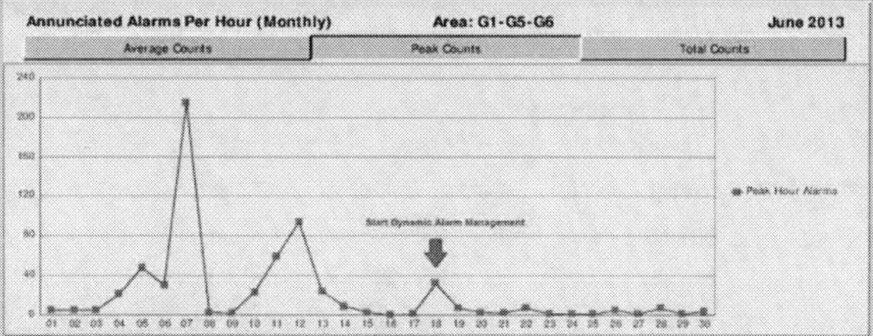

Day	Min Hour Alarms	Avg Hour Alarms	Peak Hour Alarms	Total Alarms
01-Jun-2013	0	0.3	5	6
02-Jun-2013	0	0.4	5	9
03-Jun-2013	0	0.5	5	13
04-Jun-2013	0	1.1	21	27
05-Jun-2013	0	13.3	48	319
06-Jun-2013	0	3.6	30	86
07-Jun-2013	0	11.9	215	285
08-Jun-2013	0	0.4	3	10
09-Jun-2013	0	0.3	2	7
10-Jun-2013	0	1.5	23	35
11-Jun-2013	0	4.7	59	112
12-Jun-2013	0	11.9	94	285
13-Jun-2013	0	3.8	24	90
14-Jun-2013	0	1.3	9	32
15-Jun-2013	0	0.3	3	8

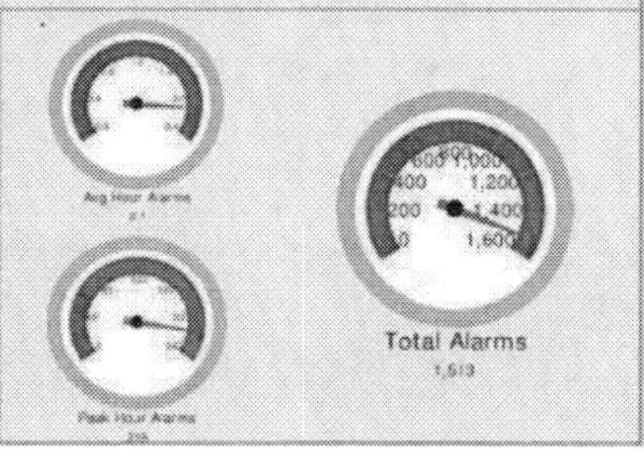

PROSYS Maximize Your Plant.™

Operator Graphics – The Risk of Changing Focus

- **Distraction of changing focus from process graphic to faceplate window**
- **Added workload from managing multiple open windows**
- **Increased probability of errors when changing values for un-intended tag with multiple faceplates open**
- **All of above become multiplied and more complex when process is transitioning from one state to another**

PROSYS Maximize Your Plant.™

Faceplate Operation

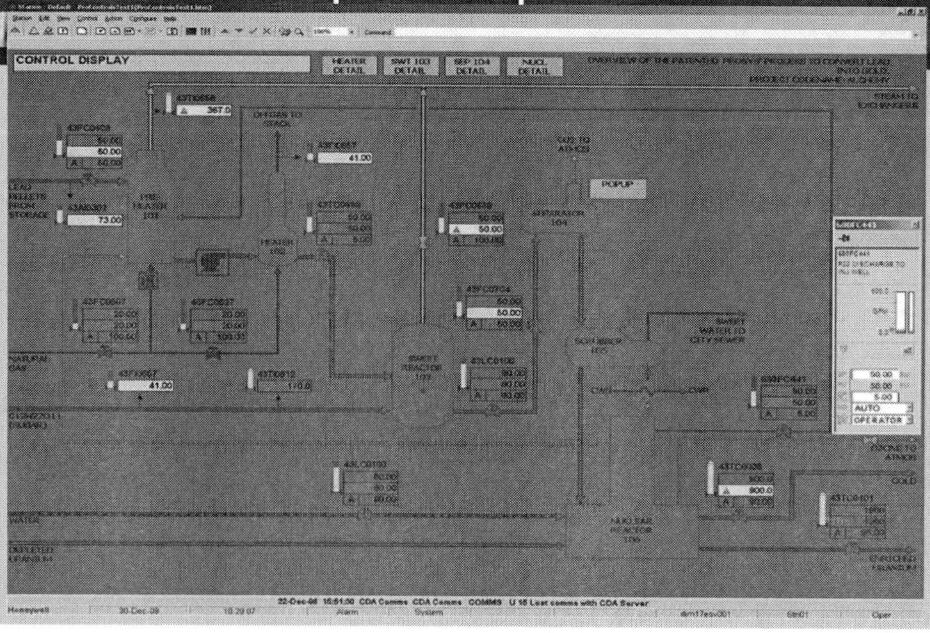

PROSYS™ Maximize Your Plant.™

Potential for Errors

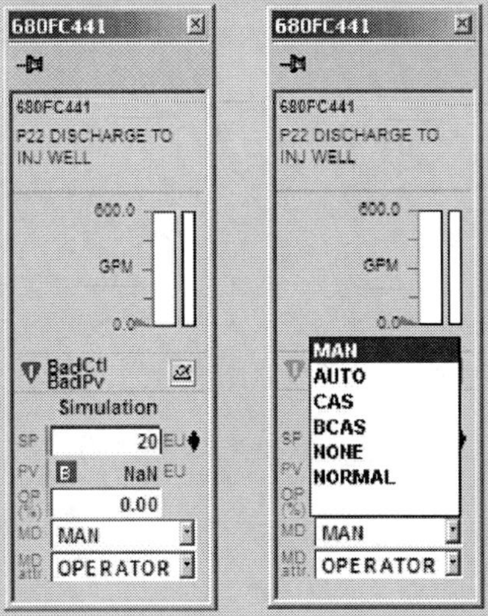

Faceplate Design -can introduce additional errors

- Mode drop down list covers SP, PV and OP values
- Operator may select wrong Mode if SP, PV and OP values are hidden from view
- Mode list offers more options than necessary

PROSYS™ Maximize Your Plant.™

Direct Entry Operation

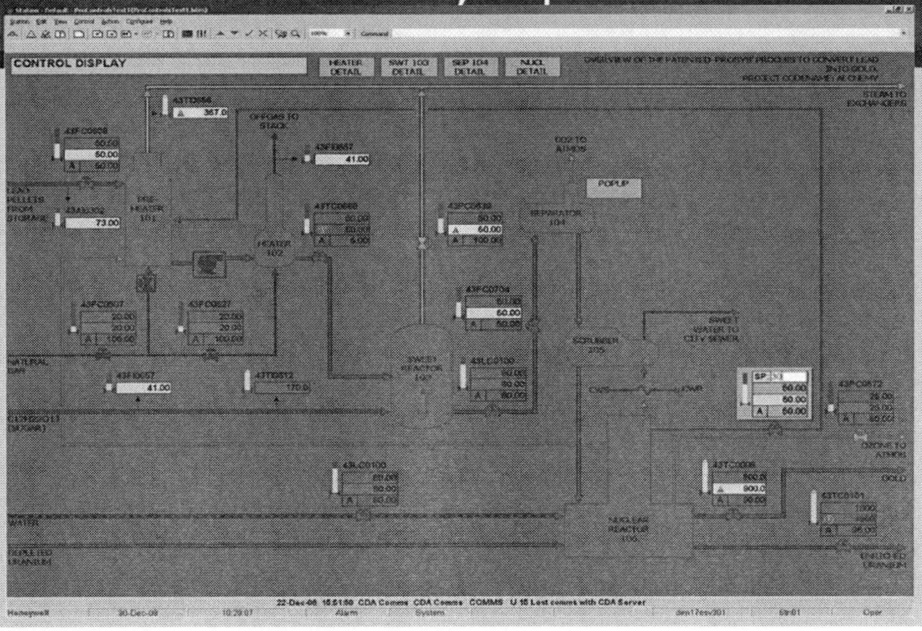

PROSYS Maximize Your Plant™

10th Global Congress on Process Safety

Direct Entry Fields

Values can be entered by:

- Select point
- Typing value via keypad
- Pressing enter

Key Factors:

- By maintaining focus of points after value change, it improves safety for input errors, multiple clicking functions and is more time efficient
- Easily recognizable operator enterable fields

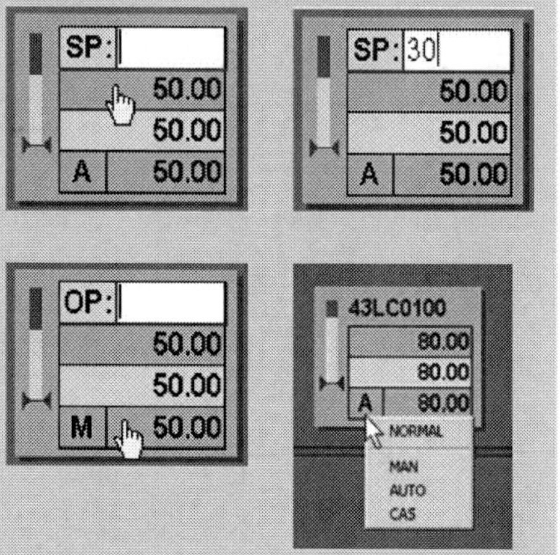

PROSYS Maximize Your Plant™

10th Global Congress on Process Safety

Evaluation of Methods

Comparison of Faceplate to Direct Entry

- **Keystroke Level Analysis**
- **Operator Loading Analysis**
- **Risk Analysis**

PROSYS Maximize Your Plant.™

10th GLOBAL CONGRESS ON PROCESS SAFETY

Keystroke Level Model

Symbol	Time (s)	Description
K	0.28	Keystroke
P	1.1	Point to object
BB	0.2	Click on object
H	0.4	Home hands on keyboard or mouse
M	1.2+	Mental act or routine thinking

PROSYS Maximize Your Plant.™

10th GLOBAL CONGRESS ON PROCESS SAFETY

Estimate Execution Time Analysis

- **Evaluates execution time by an operator comparing faceplates to direct entry for SP/OP and Mode Changes**

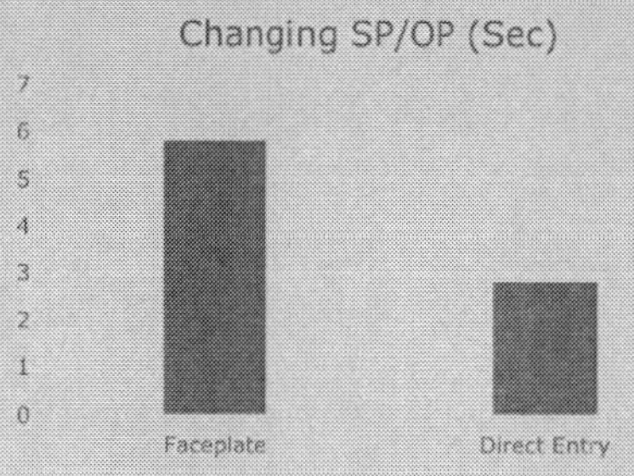

- **50 % Reduction**

PROSYS Maximize Your Plant™

10th GLOBAL CONGRESS ON PROCESS SAFETY

Estimated Operator Loading Analysis

- **Evaluates Mental and Physical Load on Operator**
  - Tasks that require very little thought are removed – such as (BB), (H)

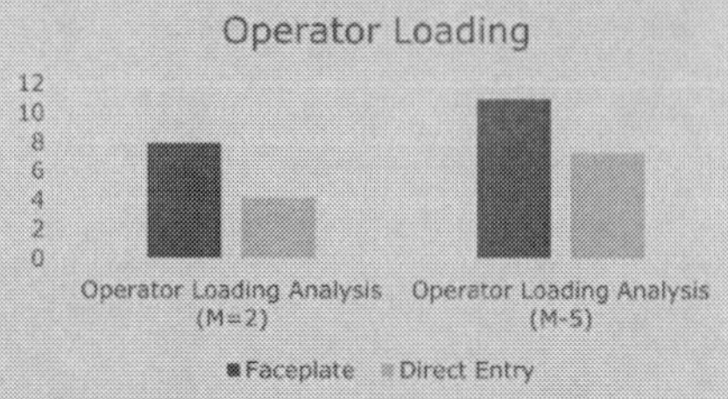

  - Mental and Physical tasks are weighted based on the amount of fatigue they cause – assigned a weight of M=2.0 to 5.0

PROSYS Maximize Your Plant™

10th GLOBAL CONGRESS ON PROCESS SAFETY

Risk Analysis

- **Identifies actions where operator entry errors can occur**

PROSYS Maximize Your Plant.™

10th GLOBAL CONGRESS ON PROCESS SAFETY

Risk Analysis

Faceplate operation KLM for changing SP/OP	Potential Error
Move to shape (P)	Select wrong point
Click on shape (BB)	
Move to faceplate (P)	Point in faceplate is previous point
Click on SP/OP field (BB)	Change wrong parameter
Move hands to keyboard (H)	
Type in value and press enter (4K)	Mistype value and press enter
Move hands to mouse (H)	
Move to close faceplate (P)	
Click to close faceplate (BB)	
Potential error count	**4**

PROSYS Maximize Your Plant.™

10th GLOBAL CONGRESS ON PROCESS SAFETY

Risk Analysis

Direct Entry operation KLM for changing SP/OP	Potential Error
Move to SP/OP (P)	Select wrong parameter or point
Click on SP/OP field (BB)	
Move hands to keyboard (H)	
Type in value and press enter (4K)	Mistype value and press enter
Potential error count	**2**

PROSYS Maximize Your Plant™

10th GLOBAL CONGRESS ON PROCESS SAFETY

Direct Entry Vs Faceplate

Summary of Benefits

- **Execution Time SP/OP Changes - 51.5% less**
- **Execution Time for Mode Changes – 50% less**
- **Operator Loading – 33 % to 47% less**
- **Risk of Errors – 50% less**

PROSYS Maximize Your Plant™

10th GLOBAL CONGRESS ON PROCESS SAFETY

Conclusion

- **Many industrial errors are a result of operators using control systems with flawed designs**
  - Poor design and performance by alarm management distracts operators and/or occludes critical alarms
  - Operator graphics with multiple open faceplates can cause a change intended for one controller to be entered into the faceplate of another
- **The two mechanisms listed above are predictable and therefore preventable through better design available today**

PROSYS™ Maximize Your Plant.

Questions?

Comments?

10TH GLOBAL CONGRESS ON PROCESS SAFETY

visium KMS
When Sharing is not Enough… HUAA
10th Global Congress on Process Safety – New Orleans, LA
Expert Insight Delivered
Osys

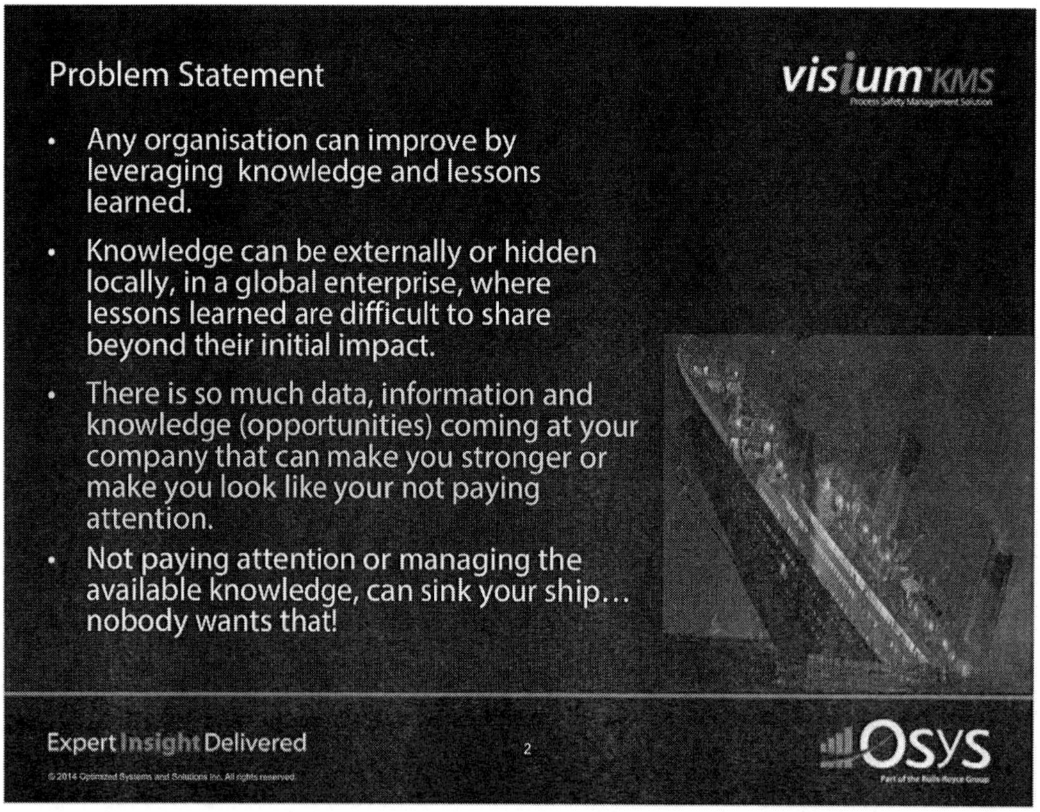
Problem Statement
visium KMS
• Any organisation can improve by leveraging knowledge and lessons learned.
• Knowledge can be externally or hidden locally, in a global enterprise, where lessons learned are difficult to share beyond their initial impact.
• There is so much data, information and knowledge (opportunities) coming at your company that can make you stronger or make you look like your not paying attention.
• Not paying attention or managing the available knowledge, can sink your ship… nobody wants that!
Expert Insight Delivered
2
Osys

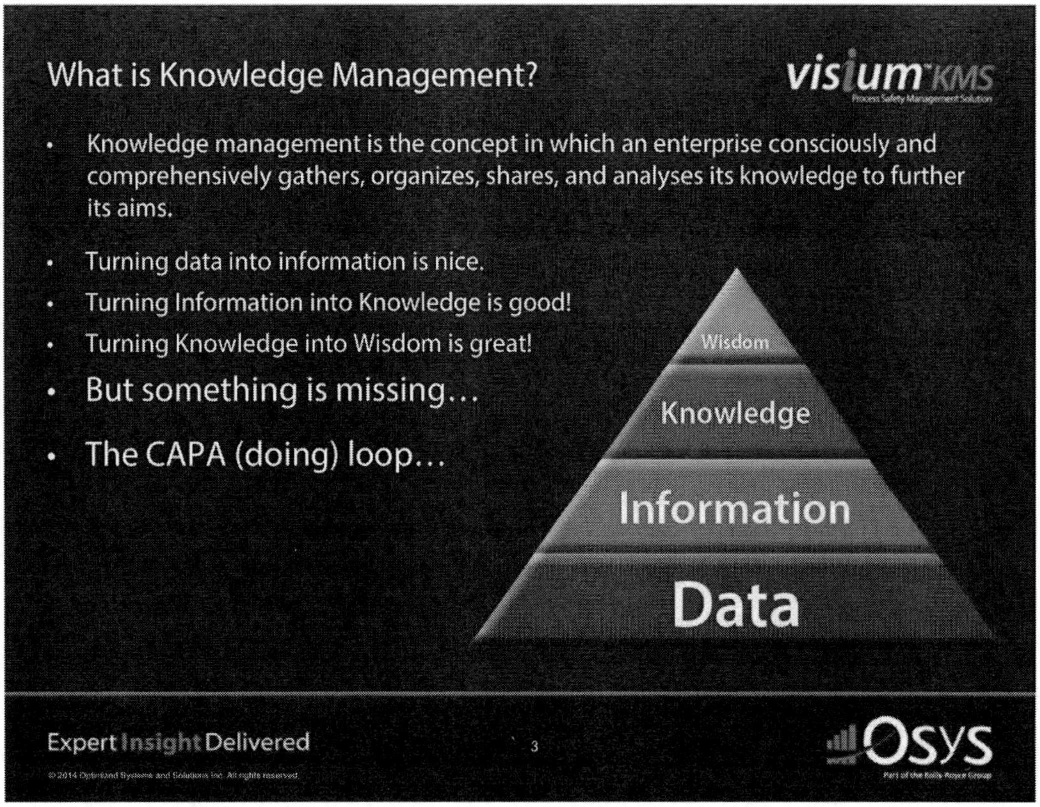
What is Knowledge Management?
visium KMS
Knowledge management is the concept in which an enterprise consciously and comprehensively gathers, organizes, shares, and analyses its knowledge to further its aims.
Turning data into information is nice.
Turning Information into Knowledge is good!
Turning Knowledge into Wisdom is great!
But something is missing…
The CAPA (doing) loop…
Wisdom
Knowledge
Information
Data
Expert Insight Delivered
3
Osys

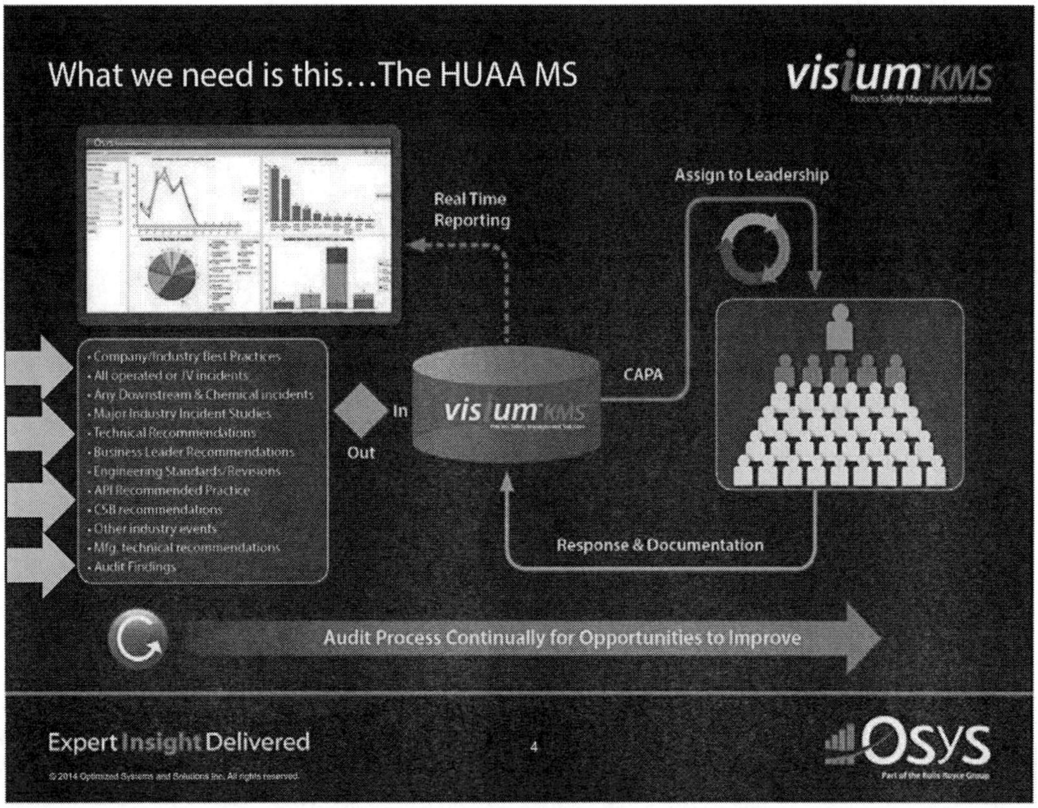
What we need is this…The HUAA MS
visium KMS
Real Time Reporting
Assign to Leadership
• Company/Industry Best Practices
• All operated or JV incidents
• Any Downstream & Chemical incidents
• Major Industry Incident Studies
• Technical Recommendations
• Business Leader Recommendations
• Engineering Standards/Revisions
• API Recommended Practice
• CSB recommendations
• Other industry events
• Mfg. technical recommendations
• Audit Findings
In
Out
visium KMS
CAPA
Response & Documentation
Audit Process Continually for Opportunities to Improve
Expert Insight Delivered
4
Osys

What is the HUAA Management System…

visium KMS
Process Safety Management Solution

HUAA – Heard, Understood, Acknowledged, Actioned.

1. Opportunities identified (heard) by your experts,
2. Captured by an electronic system,
3. Expertly Reviewed (understood),
4. Accepted or rejected (acknowledged) and
5. Assigned (actioned) to Sr. Managers for resolution and tracked to closure.

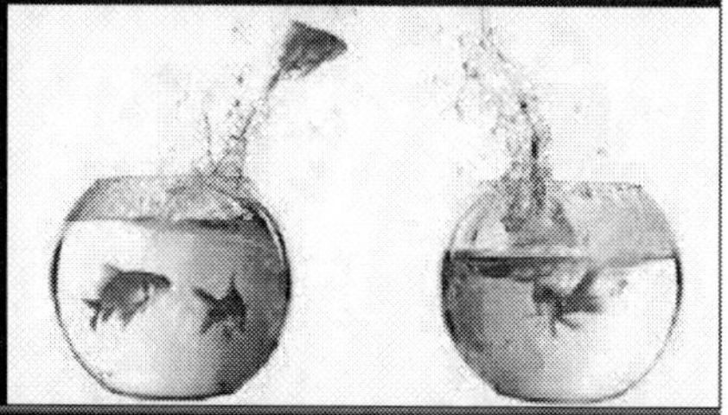

Expert Insight Delivered 5

Osys
Part of the Rolls-Royce Group

HUAA – Management Systems - Detail

visium KMS
Process Safety Management Solution

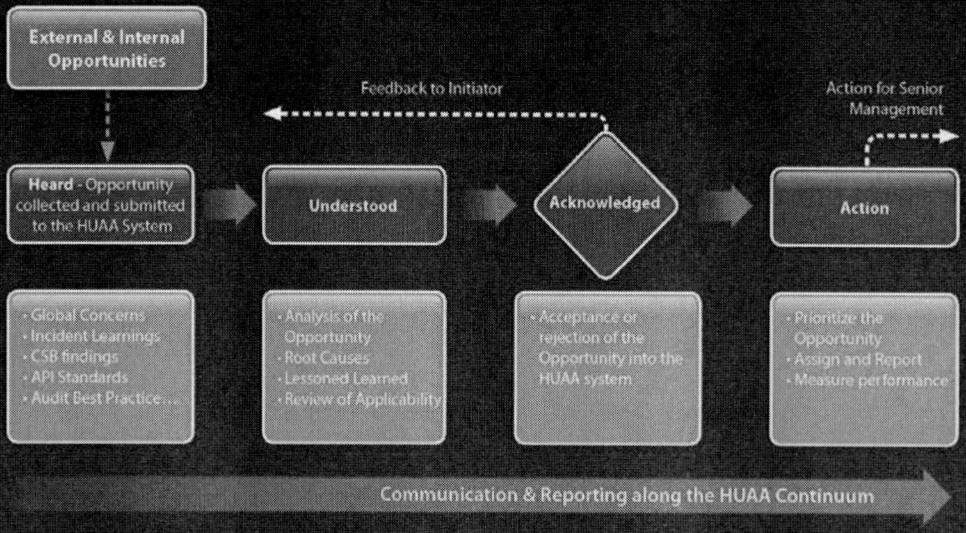

Expert Insight Delivered 6

Osys
Part of the Rolls-Royce Group

HUAA Results

visium KMS
Process Safety Management Solution

- Efficient Collection - Important Opportunities are identified and collected from many sources (internal and external), and entered into the HUAA management system.
- Expert Vetting - These opportunities are vetted, parsed and resolved for inclusion or exclusion; documenting & communicating both.
- Accountable Assignment - Included opportunities are systematically assigned to Sr. Leaders and managed across the enterprise at SVP and C levels.
- Action - Improvement will be assured through the accountable actions of these leaders.
- Results - The organization will gain competitive advantage by active listening, learning from their experience, as well as the experience of others, and by taking ACTION!

Expert Insight Delivered 7 OSyS
Part of the Rolls-Royce Group

Sharing is not enough to keep your ship afloat!

visium KMS
Process Safety Management Solution

Expert Insight Delivered 8 OSyS
Part of the Rolls-Royce Group

352631 Autocad - Smart Grid – Agriculture - Architecture and the United States Government

Abstract:

The computational methods, process control, process design, process safety, public policy of the smart chip for agriculture industry, and architecture industry in the United States utilizing the computational methods of AutoCAD in the structural development stages of the smart grid for device/component/equipment and geographical development and public policy for grant funding.

352747 Production, Characterization and Catalytic Studies of Biobased Carbon Materials

Abstract:

Carbon-based nano materials were produced from wood chips, lignin and bio-char. The influence of reaction conditions on the production of nano-carbons was investigated. Carbon-based nanomaterials were characterized using scanning electron microscopy (SEM), transmission electron microscopy (TEM), high resolution transmission electron microscopy (HRTEM), and X-ray diffraction (XRD). The catalytic performance of these nanomaterials as syngas cleaning adsorbents and chemical conversion catalysts for bio-fuel processes has been evaluated.

347835 Uncertainty in Sour Gas Viscosities Estimation, What Is the Effect on Your Reservoir Inflow and Tubing Performance

Abstract:

In light of the need to find an applicable, consistent and promising correlation to accurately calculating sour gases viscosity under several reservoir conditions, this study aims to compare the existing correlations and other industry standard methods, check their accuracy and range of applicability for sour/acidic gases. Moreover, investigating the impact of uncertainties in estimating sour gas viscosities on reservoir flow performance relationship (IPR), reserves and resources estimations, tubing performance, pipeline and facilities design, the design of compressors and more importantly the need to find such correlation to be used in further advancements of reservoir simulation.

354052 Chromonic Nanocarriers for Chemotherapeutics: Size Distribution and Control Release Studies

Abstract:

Conventional chemotherapy fails to release the drug in the controlled manner; rather they release the drug instantly due to which re-administration is required as therapeutic index falls. Moreover, they target both the cancer cells and non-cancer cells which make the treatment of the cancer cells highly ineffective due to excessive toxicities. Recently, the nanoparticle mediated drug delivery for chemotherapeutics has gained much attention. The present study reports the use of a novel class of materials – *Chromonics* in cancer drug delivery. Folic acid as a chromonic material has been explored to develop nanocarrier for sustained release of chemotherapeutic drugs *in-vitro*. With the help of a biocompatible polymer as a tool, folate nano-domains were formed from self-assembled liquid crystalline folate solutions. These nano-domains were crosslinked to form a stable nanoparticle system. The effect of different factors like crosslinking agent, relative polymer and folate concentration, on the size distribution was investigated in detail. Based on the observations, we have reported that not only the synthesis but also the size of the developed nanoparticles can be controlled by varying the folate: HPMC weight ratios. The present study then explored the release of drugs from these nanocarriers. We have investigated the release of folic acid as a model drug from these nanoparticles as well as release of methotrexate and cytarabine as an anticancer drug for the period of 60 days. It was observed that release of drug was nanoparticle's size-dependent, release medium dependent, pH of the release medium dependent, crosslinking agent and its concentration dependent. The significant part of the current study is the synthesis of chromonic nanoparticles, how the size can be controlled, how controlled release can be obtained as well as how the release profiles can be modified depending on the dosage forms. Therefore, we can say that folate nanoparticles can be a potent nanocarrier for the sustained release of cancer drugs.

Double Containment Piping Systems: A Fail-Safe, Leak-Free Solution

IPEX
Committed to Excellence

Darin Johnson, P.Eng.
Applications Engineer
IPEX USA LLC

AIChE – 2014 Spring Meeting
New Orleans, LA

Monday, March 31, 2014

Why Double Containment Piping Systems?

IPEX
Committed to Excellence

- December 1988 – law change within the ***Resource, Conversation, and Recovery Act*** in the United States regarding underground piping.
- U.S. Environmental Protection Agency (EPA) requirement: Code of Federal Regulations (CFR), Title 40, Part 280.
  - Petroleum Products
  - Hazardous Substances (over 3200, CFR Title 40 Part 302.4)
  - Hazardous Wastes

Why Double Containment Piping Systems?

- Buried tank - all buried piping connected to it must be double contained.
- Above ground tank – associated buried piping consists of 10% or more of the tank volume, it also must be double contained.

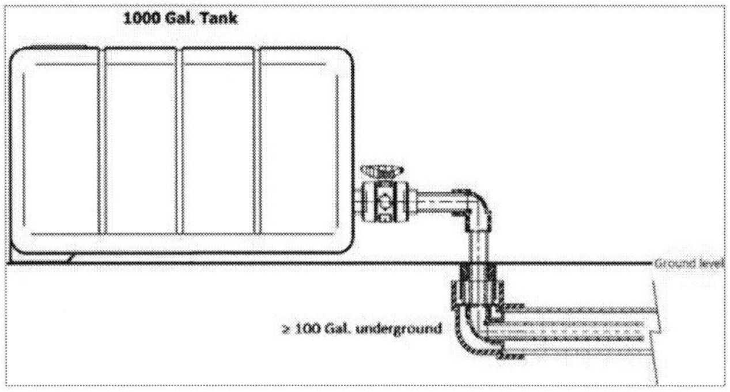

Why Double Containment Piping Systems?

- If the piping is less than 10% of the volume of the aboveground tank, and you choose not to double contain the buried piping, ***<u>responsibility for any clean-up or remediation resulting from a release will still fall on the Client.</u>***

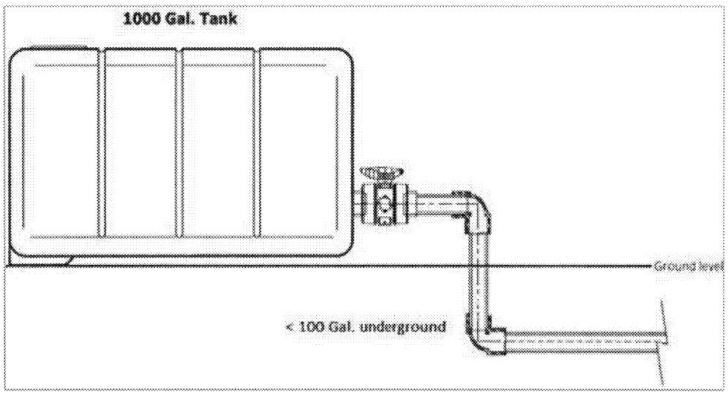

Why Double Containment Piping Systems?

- **Above Ground Piping Systems**
  - Different jurisdictions have to ability to require chemicals be double contained.

Nomenclature of Double Containment Piping

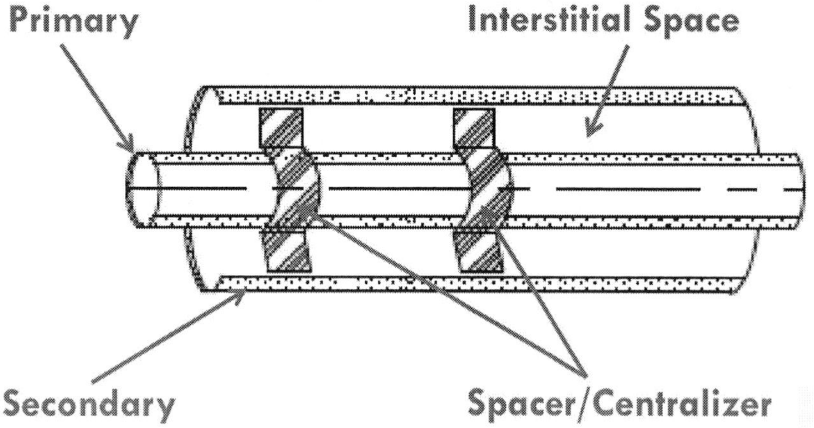

Double Containment Piping Systems: Material Combinations

SYSTEM Carrier	Containment	Carrier SIZE inches
PVC 80	PVC 80	1/2 - 12
PVC 80	PVC 40	1/2 - 12
PVC 40	PVC 40	1/2 - 12
PVC DWV	PVC DWV	1-1/2 - 12
PVC 80	CS SCH 10	1/2 - 12
CPVC 80	CPVC 80	1/2 - 8
CPVC 80	PVC 80	1/2 - 12
CPVC 80	PVC 40	1/2 - 12
CPVC 80	CS SCH 10	1/2 - 12
NAT PP 80	PVC 80	1/2 - 2
NAT PP 80	PVC 40	1/2 - 2
RED PVDF 80	CPVC 80	1/2 - 2
RED PVDF 80	PVC 80	1/2 - 2
NAT PVDF 80	PVC 80	1/2 - 2
ENFUSION PP	PVC DWV	1/2 - 6
S40 A53B CS BW	PVC 80	1/2 - 8
S40 A106B CS BW	PVC 80	1/2 - 8

SYSTEM Carrier	Containment	Carrier SIZE inches
S40 A53 BLK SW	FRP	1/2 - 2
S40 A53 BLK SW	PVC 80	1/2 - 2
S80 A53B CS SW	PVC 80	1/2 - 2
S10 T304L SS BW	PVC 80	1/2 - 8
S10 T304L SS BW	PVC 40	1/2 - 8
S10 T0304L SS BW	CPVC 80	1/2 - 8
S10 T304L SS BW	FRP	1/2 - 12
S10 T304L SS BW	S10 T304L	1/2 - 12
S40 T316L SS BW	PVC 80	1/2 - 8
S40 T316L SS BW	CPVC 80	1/2 - 8
S40 T316L SS BW	FRP	1/2 - 12
S40 T304L SS SW	PVC 80	1/2 - 2
S40 T304L SS SW	PVC 40	1/2 - 2
S40 T304L SS SW	CPVC 80	1/2 - 2
S40 T304L SS SW	FRP	1/2 - 2
S10 T316L SS SW	PVC 80	1/2 - 2
S10 T316L SS SW	CPVC 80	1/2 - 2

Double Containment Piping Systems: Design Considerations

- **Material Selection**
  - Primary Pipe
    - Chemical Compatibility
    - Pressure Capabilities
    - Thermal Properties
  - Secondary Pipe
    - Chemical Compatibility
    - Loading (Static & Live) for buried applications
    - Hanging requirements for above ground
    - Leak detection requirements
    - UV requirements

Double Containment Piping Systems: Design Considerations

- **Thermal Expansion & Contraction**
  - Two basic thermal relationships may be present:
    - When only one of the two systems experiences thermal expansion/contraction.
    - When both systems experience different magnitudes of thermal expansion/contraction.
  - The Floating System
  - The Restrained System

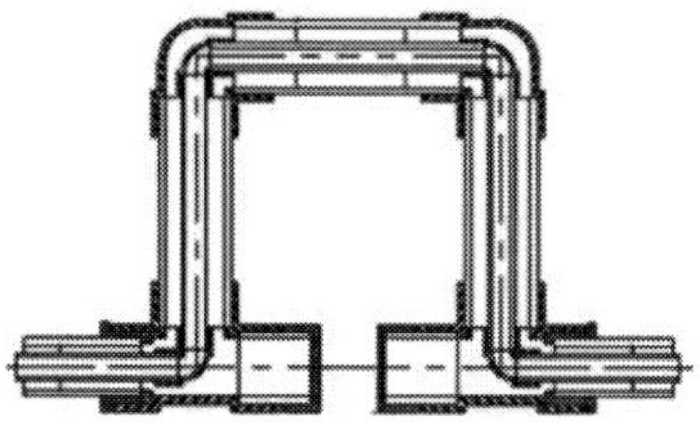

Double Containment Piping Systems: Leak Detection Requirements

- EPA requirement: (CFR), Title 40, Part 280
  - Pressurized systems require electronic leak detection that must be checked once per year.
  - Drainage systems (gravity flow) require visual leak detection with monthly manual inspection for a media leak.

Double Containment Piping Systems: Leak Detection Requirements

- The following forms of leak detection are acceptable as per regulations
  - Visual (drainage applications only)
  - Electronic low point probes
  - Continuous cable

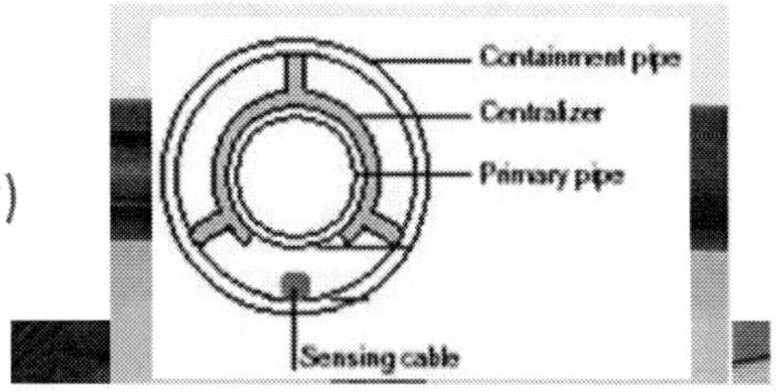

Double Containment Piping Systems: Applications

- Underground Applications
- WTP & WWTP
- Oil & Gas
- Hospitals
- Museums
- Food and Beverage
- Pharmaceuticals
- Laboratories
- Data Centers

Double Containment Piping Systems: A Fail-Safe, Leak-Free Solution

Questions?

Thank you!

DTP Process

"Now Ready to License"

JGC Corporation
Mitsuo Morita

2014

JGC CORPORATION | MITSUBISHI CHEMICAL

What is DTP Process ?

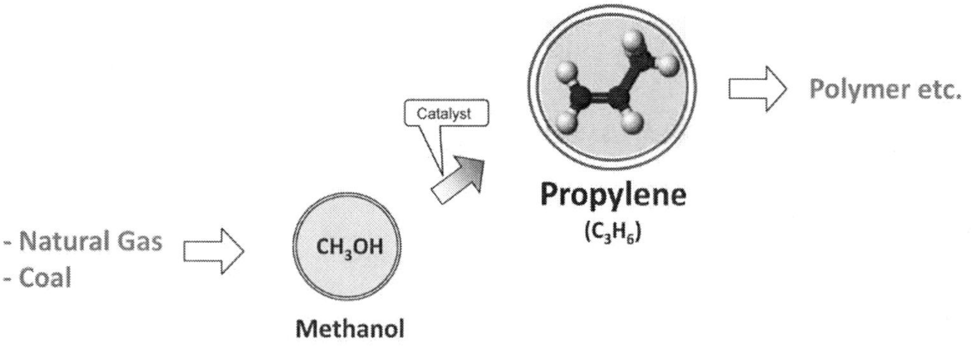

-On-Purpose Propylene Production from Methanol

JGC CORPORATION - 2 - MITSUBISHI CHEMICAL

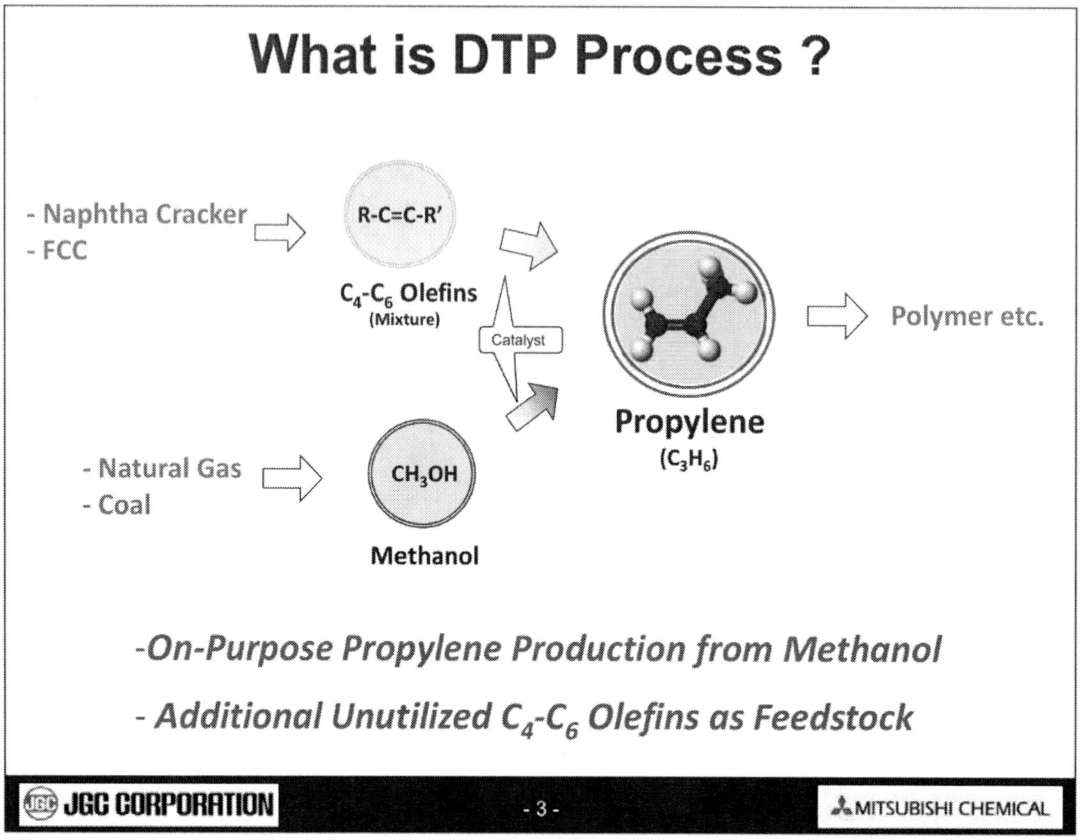
What is DTP Process ?
- Naphtha Cracker
- FCC
R-C=C-R'
C4-C6 Olefins
(Mixture)
Catalyst
Polymer etc.
Propylene
(C3H6)
- Natural Gas
- Coal
CH3OH
Methanol
-On-Purpose Propylene Production from Methanol
- Additional Unutilized C4-C6 Olefins as Feedstock
JGC CORPORATION
- 3 -
MITSUBISHI CHEMICAL

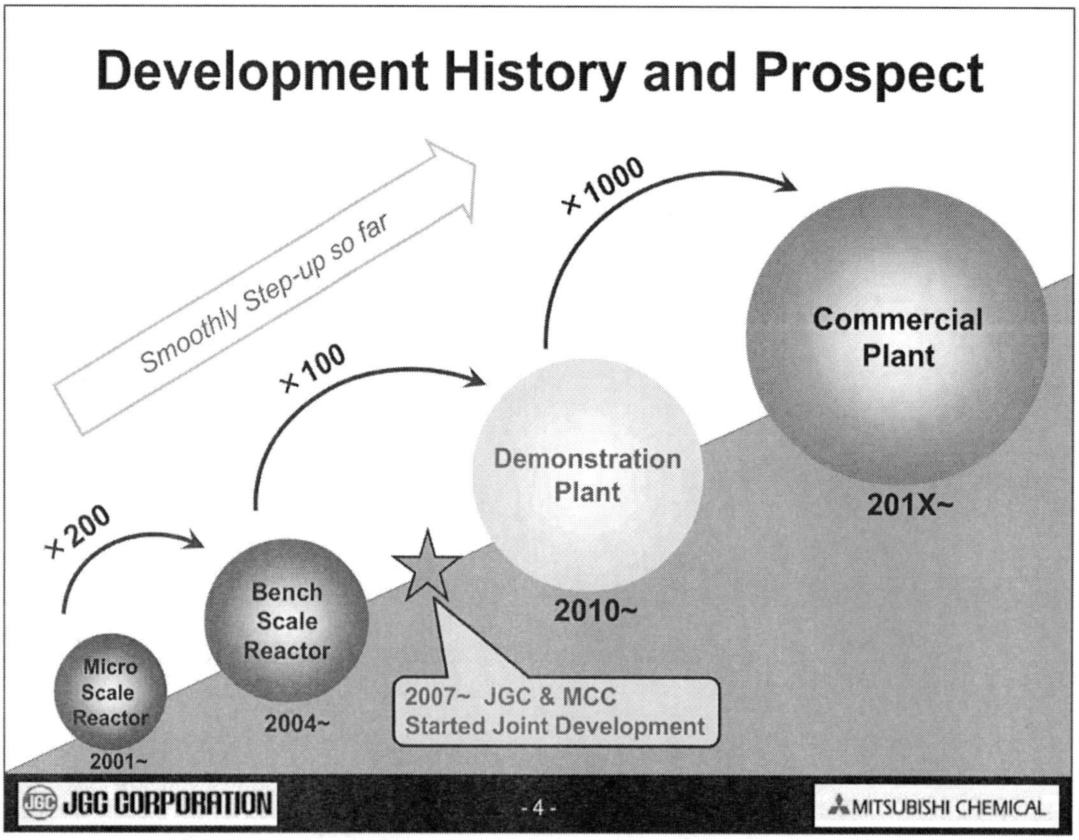
Development History and Prospect
Smoothly Step-up so far
×1000
×100
×200
Commercial Plant
201X~
Demonstration Plant
2010~
Bench Scale Reactor
2004~
Micro Scale Reactor
2001~
2007~ JGC & MCC
Started Joint Development
JGC CORPORATION
- 4 -
MITSUBISHI CHEMICAL

Demonstration Plant Overview

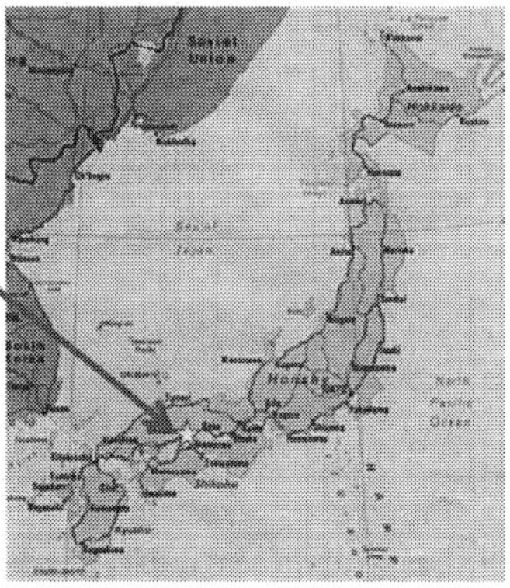

- **Located in Mitsubishi Chemical Mizushima Plant**
- **Successfully Start-up in August 2010**
- **Continuous Operation for Two Years**

Process Flow (Standalone Case)

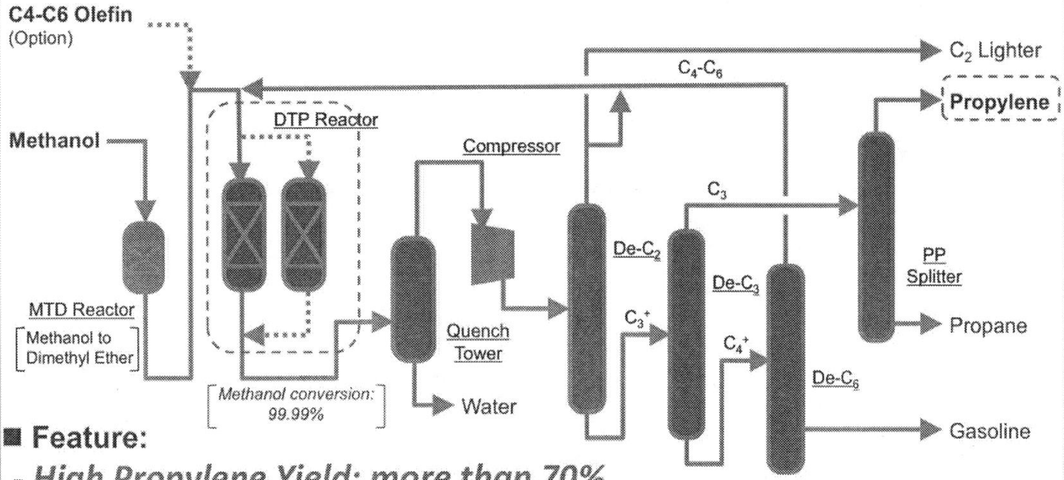

■ **Feature:**

- *High Propylene Yield: more than 70%*
- *Applied Existing Proven Technologies*

Process Flow (Integration Case)

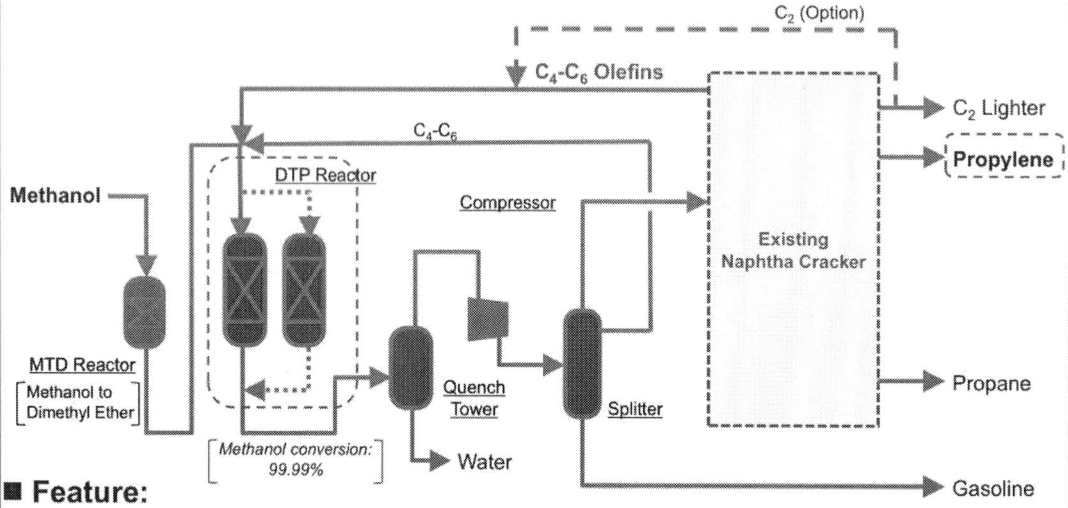

■ **Feature:**

- ***High Propylene Yield: more than 70%***
- ***Applied Existing Proven Technologies***
- ***Low Investment Cost in case of Integration with Naphtha Cracker***

JGC CORPORATION - 7 - MITSUBISHI CHEMICAL

Application

(1) Naphtha Cracker Integration Case (Propylene Capacity: 20KTA – 200KTA)

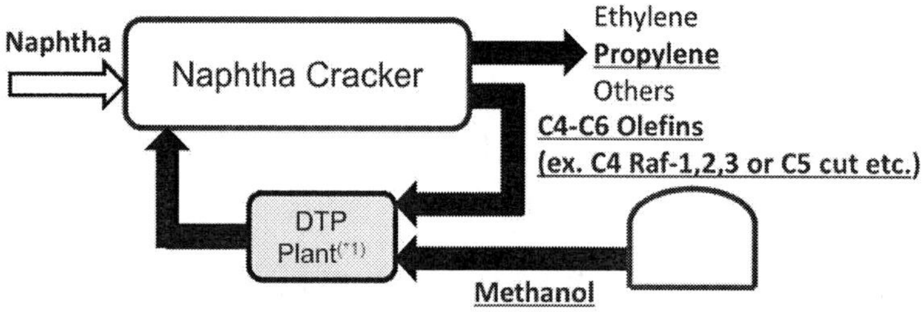

(*1) Utilizing facilities in Naphtha Cracker.

(2) Standalone Case (Propylene Capacity: 250KTA – 500KTA)

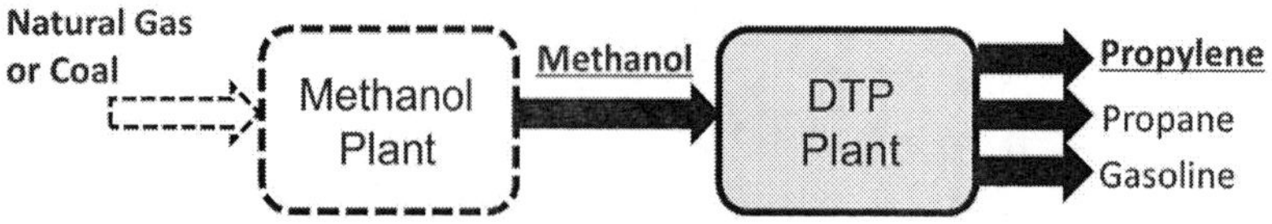

- ***Two Cases available depending on Client requirements !!***

JGC CORPORATION - 8 - MITSUBISHI CHEMICAL

DTP is…

- *High Yield Propylene Production Process*
- *Flexible Feed Methanol to Olefin Process*
- *Process that increase P/E Ratio for Cracker*

Thank you very much !!

Contact:
Mitsuo Morita
Market Development Dept.
Global Marketing Division
2-3-1, Minato Mirai, Nishi-ku,
Yokohama 220-6001, Japan

Phone: 81-45-682-8371
FAX: 81-45-682-8183
E-Mail: morita.mitsuo@jgc.co.jp

JGC CORPORATION - 9 - MITSUBISHI CHEMICAL

352533 Self-Cleaning "Bernoulli" Type Filters Used in Onshore and Offshore Applications

Abstract:

Georg Schuenemann GmbH (SAB-BREMEN) has developed a new Filter Type F-480, dubbed the "Compact Giant" for applications in onshore water intake and cooling water systems, sea water desalination plants, and in offshore projects, such as FPSO's, Drillships, and Production Platforms.

This paper will describe the essential design features of the filters, based on the "Bernoulli principle" which result in a fully automated self-cleaning filter system, and choice of materials in a sea water service with high corrosion rates.

A case study will be presented in which stringent project requirements such as substantial flow rates, adaptibility to space limitations and pipe connections had to be observed. Reliability of the system, extended run times without the need for frequent replacement of filter elements, as well as ease of monitoring and operator involvement have been achieved.

352653 Effects of Support on Sulfur Tolerance and Regeneration of Pt Catalysts Measured By Ethylene Hydrogenation and EXAFS

Abstract:

We investigated the effect of support on sulfur tolerance and regenerability under reducing environments by measuring the rate of ethylene hydrogenation, combined with hydrogen chemisorption, and extended x-ray absorption fine structure (EXAFS). Alumina and titania (P25) supports were used to prepare Pt catalysts at 1 wt.%. The catalysts were tested after treatment in H_2S/H_2 at 250 °C followed by regeneration treatments in H_2 at 250 °C, 350 °C and 450 °C. The sulfided catalysts exhibited a 20-27 times decrease in the rate of ethylene hydrogenation and a 4-6 times drop in the Pt surface area. Regenerations up to 450 °C were unable to remove all the sulfur, as shown by the presence of Pt-S bonds by EXAFS. However, a partial recovery of the hydrogenation rate per mole of Pt was observed on sulfided Pt/Al_2O_3 after reduction at 450 °C, while the induction of strong metal support interactions (SMSI) at reduction temperature above 350 °C was observed on Pt/P25, regardless of the presence of sulfur. For Pt/P25, the reversal of the SMSI state together with sulfur removal by mild oxidation suggests that sequential reduction/oxidation treatments may be more effective in restoring the S-free state of TiO_2-supported catalysts.

352799 Improved Operational Efficiency and Reliability through Insulation Materials Selection

Abstract:

This paper will explore the implications of insulation materials selection on the energy efficiency and reliability for Ethylene production units. Considerations will include insulation properties, structural support, installation methods, installation and maintenance costs.

BOILER MACT COMPLIANCE: MERCURY IN FUEL GAS

PATRICK LAINE, WES YONGE, & VIC VICKERY
PEI, HOUSTON, TX 77043

Patrick Laine, Ph.D.
Chief Scientist
Houston, TX
850-866-6686
plaine@pei-tx.com

Wes Yonge, PE
Chief Engineer
Houston, TX
713-806-8396
wyonge@pei-tx.com

James 'Vic' Vickery
Chief Technology Officer
Houston, TX
281-536-0899
vvickery@pei-tx.com

1 Introduction & Objectives

For most ethylene producers the primary objective of this sampling and analysis effort will be to verify and "establish a mercury fuel specification through which gas-fired boilers that use a fuel other than natural gas or refinery gas may be considered Gas 1 unit" as identified in 40 CFR Part 63.7521.(f).

Measurement and Monitoring Requirements: Decision Rules

❑ If Hg concentration ≤ 50% of Gas 1 specification* then no further testing required

❑ If Hg concentration > 50% and ≤ 75% of the specification* then semi-annual sampling required

❑ If Hg concentration > 75% of the specification* then monthly sampling required

*Specification is **40 µg/Sm³**

The goal will be to execute a sampling and analysis campaign that will produce compliance quality data. The key to producing and validating compliance quality data is a well-designed and executed QA/QC protocol. At a minimum, a performance based QA/QC protocol is recommended that includes the following components:

- Sorbent Trap Breakthrough Bed:
  - ❑ Sorbent traps consisting of a minimum of two sorbent beds in series so that the trailing bed can be analyzed to verify that no breakthrough occurred from the leading bed.
- Duplicate Samples:
  - ❑ The sampling system should be configured with two trains to accommodate the collection of simultaneous duplicate sorbent trap samples, in parallel, providing a means of verifying precision in the sample collection process.
- NIST (National Institute of Standards and Technology) Traceable Spike:
  - ❑ Sorbent traps with a third bed (downstream) that is spiked with a NIST traceable spiked mass of mercury serves as a matrix spike and provides a means of verifying that components in the matrix (the fuel gas) are not affecting the trapping efficiency of the sorbent trap.

2 Methods

Sorbent Trap Mercury Sampling Methods Comparison

Method	Description	Method Type: S = Sampling A = Analysis	Evaluation	Sample Collection Phase QA/QC
ISO 6978	Hg in Natural Gas by Thermal Desorption (CVAFS)	S & A	Method calls for low sample volumes can lead to non representative results.	No
EPA Method 30B	Determination of Gaseous Hg Emissions from Coal Fired Combustion Sources using Carbon Sorbent Traps	S	Requires modification (Modified M30B)	Yes
ASTM D6350	Hg S&A in Natural Gas (AFS)	S & A	Method calls for short term samples (i.e., low sample volume) which leads to poor accuracy.	No
ASTM D5954	Hg S&A in Natural Gas (AAS)	S & A	AAS is less sensitive than AFS.	No
EPA Method 1631	Measurement of Hg in Water	A	No	Yes

3 EPA Method 30B QA/QC Performance Criteria

QA/QC Test or Specification	Acceptance Criteria	Frequency
Analyzer calibration	Correlation Coefficient: 0.999	Daily
Transfer efficiency of analyzer	%Transfer Efficiency: 90%–110%	Once per sorbent trap
Initial calibration blank	< 0.5 ng	Once before each calibration
Sorbent Trap Section B breakthrough	≤ 10% of Section A Hg mass, for Hg concentrations > 1µg/dscm ≤ 20% of Section A Hg mass, for Hg concentrations ≤ 1µg/dscm	At each sample point
Spike Recovery Test	Average recovery between 75% and 125 % for Hg_0	Once per sample point
Paired Sorbent Traps Agreement	≤ 10% RD for Hg concentration > 1µg/dscm ≤ 20% RD or 0.2 µg/dscm absolute difference for Hg concentrations ≤ 1µg/dscm	At each sample point

4 Equipment

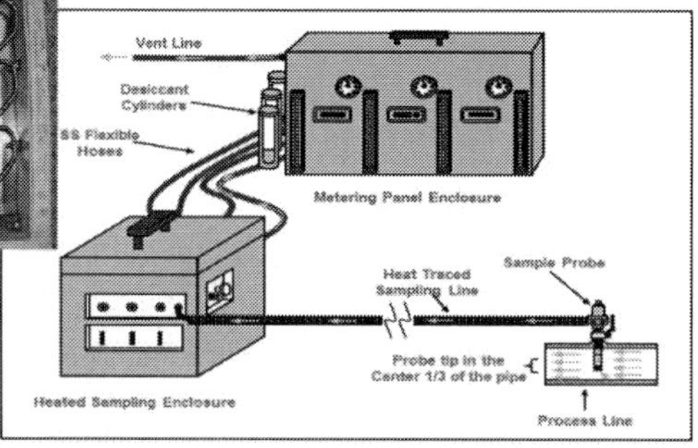

Flow

QA/QC Acceptance Criteria:	99% of Total Collected Hg	≤ 10% of Primary Mass	Recovery between 75% and 125%

Gold Trap:

5 Conclusions/PEI Recomendations

❑ There are five methods accepted by the Boiler MACT rule. PEI has extensive experience and expertise employing **all five** of these methods.

❑ Operators can achieve compliance without submitting a Fuel Analysis Plan by using ASTM D6350, however, we have noticed variable results with this method.

❑ The **PEI recommendation** for obtaining the most accurate compliance quality data is to use Modified EPA Method 30B in conjunction with EPA 1631 where traps should be analyzed in a certified ISO 17025 laboratory. The recommended pathway includes a rigorous, performance based QA/QC protocol that validates accuracy and precision of all sample collection procedures.

351376 Young Professional Simulation Tutorial

Abstract:

18C00 Young Professional Simulation Tutorial

Authors: Naomi Hua and Mike Donahue, Invensys

Process simulation is a complex skill to master. Even with years of experience, a typical user is constantly facing new problems that need to be addressed. There are many issues that can lead to unexpected and unnecessary errors with the use of process simulation tools. Regardless of what simulation program you are using, these issues are common from one program to another. In this presentation, targeted for the young professional, we will discuss the "Most Common Pitfalls in Process Simulation". For each topic, we will discuss various techniques and methods these troublesome problems and prevent their occurrence.

Topics to be discussed include:

1. Erroneous data entry

2. Simulation defaults

3. Bad plant data

4. Thermodynamic method selection

5. Simulation tolerances and convergence

6. Recycle identification and handling

7. Model complexity

8. Difficult or impossible specifications

9. Assay pseudo-component development

10. Assay property blending

1. Erroneous data entry is the most common mistake made when building a process simulation model. This presentation will cover methods that can help ensure proper data entry in your process simulation model, such as checking the data you entered, particularly the units of measure, using "alternate forms" of data input such as copy/paste and data import, and reviewing all warning and error messages.

2. Defaults are methods, values, and conditions that are automatically selected for you by the program when you begin a new simulation model. These defaults will affect the results calculated by your model and every simulation program has them. Most defaults within simulation programs fall under categories such as convergence criteria, thermodynamic methods, tolerance and units of measure. The problem with defaults is unintended results from not fully understanding the effects a particular default has on the model. Take the time to understand the defaults of the program and how these defaults will affect the results. If you do not agree with the defaults selected by the program, most simulation programs will allow the user to change the default.

3. Plant data changes over time. The feed information and process conditions of the unit operations may be different. Not only does plant data change over time, but measurement information is often very inaccurate. This could be due to faulty or poorly calibrated measurement equipment, leaks in lines, sensor failures, and other issues. Because of these changes and inaccuracies over time, you always have to validate the information that is entered in your model to ensure that you are obtaining the proper results. To do so, you can utilize programs that can help with data validation and reconciliation like SimSci's ROMeo online optimization program. With it, you can connect your model to your data historian to constantly import current data measurements into your model and then reconcile these measurements to determine when problems arise within your plant and where those problems are occurring. You will also need to use your engineering judgment to determine if the data is valid. More details will be provided during the presentation.

4. The proper selection of thermodynamic methods for your simulation is the most important decision that you make when modeling a process. The method you select could have a drastic effect upon your results. Often, any number of thermodynamic methods is suitable for a given application. It is up to the user to determine the best method to use based upon knowledge, personal experience, and best judgment. Some important factors that need to be considered are the components being used and the operating conditions.

Oftentimes, the selected thermodynamic method still needs to have additional inputs such as binary interaction parameters to simulate the process accurately.

5. Many of the calculations in process simulation are iterative. When these calculations iterate, the program compares the results calculated from one iteration to the results calculated in the next iteration in order to determine

whether it is converged to a solution or needs to continue calculating. This method of comparing results is where tolerances come into play and simulation programs use tolerances to determine when a solution is reached. One of the major mistakes when modeling a process is not understanding or using the correct tolerance values. A looser tolerance may make the file easier and quicker to converge, but it also reduces the accuracy of the model. A tighter tolerance may be more accurate, but might make it impossible for the model to solve or increase the convergence time exponentially. A common practice is that when you have reached a converged solution with your model, tighten the tolerances within the model and solve it again. If you notice that your answers have changed significantly, you need to continue tightening the tolerance values until your results change little from one iteration to another. You will then know that you have reached an accurate solution for your model.

6. Recycles in process simulation occur when you return to a previously calculated unit operation to recalculate the unit due to a change made elsewhere in the model. They are very common and can often be difficult to converge. Some suggestions on troubleshooting recycle convergence include initially solving your model without the recycles to see how it behaves, providing initial estimates for recycle streams, using tighter tolerances on unit operations within recycle loops than the total recycle tolerances themselves, and ensuring that all components have a way in and out of a recycle by performing mass balances and using purge or makeup streams where necessary.

7. The general rule of thumb when modeling is to start simple and add complexity as you go. Each time you add a few streams or unit operations converge and save the model before moving on. Make sure you justify any additions that you make to the model. You can most likely model your system without including every unit from your PFD in the simulation. When modeling, think about the goal of your simulation and not about literally translating every real-world piece of equipment into the simulation model. Make the judgment call on what is most important and what the purpose is of every unit you add to your simulation. For example, if your real-life process shows a heat exchanger, a pump and a splitter, but all that is necessary is that the temperature, pressure, and separation are achieved properly, consider using a flash drum to accomplish all three.

8. Specifications within process simulation are everywhere and every stream and unit operation you place in your model contains specifications. In its simplest form, a specification is a constraint that you place on the model that the model needs to meet in order to successfully converge. For streams, these are usually temperatures, pressures, flowrates and compositions. For unit operations, they range from process conditions, like temperatures and pressures, to product conditions, like an overhead composition in a column. There are some pitfalls, however, that you can run into with specifications such as azeotropes in columns, high purity specifications, and the possibility of multiple solutions.

9. Petroleum assays are sets of distillation data that are used to characterize all of the components that reside in a given crude sample. When you enter distillation data into a process simulator, the program converts it into components that it can use when modeling your process. The components that are generated are called pseudo-components, and each pseudo-component represents all of the unknown components that would be within a given boiling point range in the crude. Some of the things you will want to ensure when properly developing your assay components are did you use the correct distillation curve basis, did you enter the data correctly, did you use enough cuts and are you using the correct characterization methods .

10. When you have more than 1 assay in a simulation model, you need to be concerned with assay blending and how it will affect the results of your file. Assay blending is the mixing of 2 or more groups of components (i.e. "C sets of cuts/pseudo-components) in order to calculate averaged component properties for cuts that are in the same boiling point range as one another. Once these new averaged component properties are determined, the program will use these averages from this point on in the simulation model any time it needs to access component

properties for a given pseudo-component. All component properties are blended, including normal boiling points, specific gravity, molecular weight and refinery inspection properties, and usually the blended component properties are calculated using a weighted average

These topics are based on experience and feedback from end-users to technical support representatives for process simulators. They often represent the most commonly asked questions or areas that most users find to be problematic. While the specific examples highlighted in each topic come from SimSci software, the intent of this tutorial presentation is to be broad-based and is particularly applicable to young professionals seeking to develop their experience with simulation.

Polymeric Sulfonic Acid Composites for Fuel Cell Applications: Effects of Casting Solvent and Sonication Time on Membrane Performance

Jimoh K. Adewole[1], Abdullah S. Sultan[1,], Amir Al-Ahmed[2], and S. M. Javid Zaidi[3]*

[1]Center for Petroleum & Minerals, King Fahd University of Petroleum & Minerals, Dhahran 31261, Saudi Arabia

[2]Center of Research Excellence in Renewable Energy, King Fahd University of Petroleum & Minerals, Dhahran 31261, Saudi Arabia

[3]School of Chemical Engineering, The University of Queensland, Australia

Abstract

Polymer blends comprising sulfonated poly (ether ether ketone) (SPEEK) and 3wt% (2-acrylamido-2-methyl-1-propanesulfonic acid) (PAMPS) were prepared by solution casting method. Two solvents (N,N-Dimethylacetamide (DMAc) and N-Methyl-2-pyrrolidone (NMP)) were uesd separately. Residual solvents in the membrane samples after drying were quantified using thermo gravimetric analysis. Effects of periods of sonication (0 to 90 minutes) and types of solvent on the membrane performance were evaluated interms of morphology and water uptake. An attempt was made to correlate membrane performance with the time of sonication in order to select the most suitable solvent and membrane casting conditions for subsequent membrane preparations. For both solvents, the water uptake was observed to first decrease at 10 minutes of sonication and then increased as the sonication time was increased. The minimum water uptake recorded for DMAc samples was 19.23% at 20 minutes sonication time while that of NMP sample was 24.24% at 10 minutes. However membrane morphology was found to deteriorate with increase in sonication time. No correlation was obtained between the membrane properties and the sonication time. In general, membrane performance was better for membranes samples using DMAc than NMP solvent. Therefore, DMAc was selected as the casting solvent for membrane preparation on which further studies will be carried out.

Keywords: morphology, water uptake, residual solvent, thermogravimetric analysis

Introduction

Water soluble polymers have received academic and industrial interests due to their potential applications in fuel cells and other electrochemical devices (Ye et al. 2012). One of the water soluble polymers that have attracted publication interest is poly (2-acrylamido-2-methyl-1-propanesulfonic acid) (PAMPS) (Florjańczyk et al. 2003). Generally, some of the major performance criteria for fuel cell materials include proton conductivity, water and methanol uptake, good mechanical properties, chemical stability, thermal stability and methanol permeability (Gil et al. 2004; Swier et al. 2005; Ye et al. 2012). PAMPS was reported to posses higher proton conductivity than partially hydrated Nafion. In addition, it has been found to be more tolerant to water fluctuation in high temperature operations. Some of the major drawbacks in the use of PAMPS are its poor mechanical properties, thermal stability, water and methanol uptake and solubility in many solvents (Florjańczyk et al. 2003). One of the approaches that have been proposed to solve these problems is development of polymer composites.

This research paper investigates the synergistic effect of combining water soluble PAMPS with sulfonated poly (ether ether ketone) (SPEEK) which is non - water soluble polymer to prepare composite membranes for application in fuel cells. Specifically, preliminary results on the performnance of the composite were presented. Performance were evaluated based on the composite morphology and water uptake. Effects of sonication time and types of casting solvent on membrane properties were also evaluated.

Material and Method

Material

Sulfonated poly (ether ether ketone) (SPEEK) purcahsed from FuMA-Tech, Germany, and poly (2-acrylamido-2-methyl-1-propanesulfonic acid) (PAMPS) purchased from Aldrich Chemical Co., are the two major polymers used in this research. The SPEEK has a DS of 0.533 and solubility of 99% in DMAc. SPEEK is derived by sulfonation of Poly (ether ether ketone) (PEEK). PEEK is a commercially successful polymer due to its excellent thermal stability and solvent resistance (Mohr et al. 1991). For sometimes, further application of PEEK have been stalled due this solubility issues. By sulfonation, PEEK can be made soluble in most commercial solvents. Importantly, the solubility of SPEEK also depends on its degree of sulfonation (DS). Since, this work involves the use of two polymers, a solvent which will disolve both the PAMPS and SPEEK must be selected. Literature review showed that PAMPS dissolves in N,N-Dimethylformamide (DMF), N,N-Dimethylacetamide (DMAc), propylene carbonate (PC), and water (Florjańczyk et al. 2003). Initial solubility studies that was carried out in our laboratory was done for N,N-Dimethylacetamide (DMAc) and N-Methyl-2-pyrrolidone (NMP). The choice of NMP was due its unique solvent properties and its structtural distiction from DMF and DMAc (Figure 1). Moreover, no literature was found where NMP was used as casting solvent for PAMPS. Thus, DMAc and NMP were used as solvents in this work. All materials were used as received from the manufacturers.

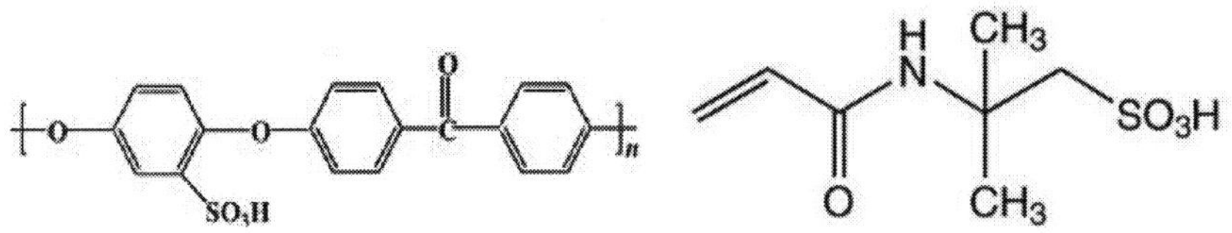

Figure 1 Structure of SPEEK and PAMPS

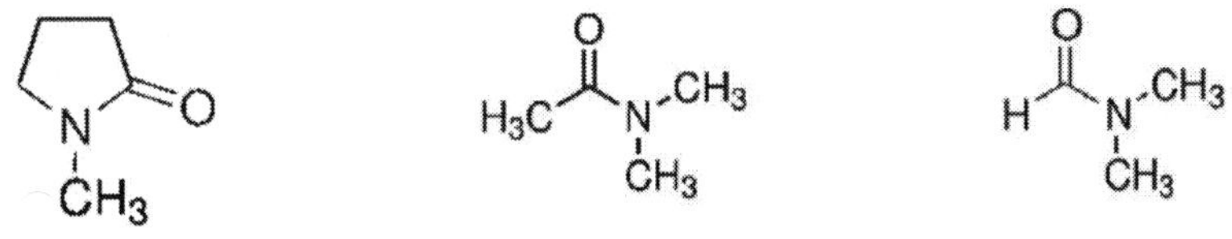

N-Methyl-2-pyrrolidone N,N-Dimethylacetamide N,N-Dimethylformamide

Figure 2 Structure of Solvents

Experimental Procedure

Membrane Preparation

SPEEK was first dissolved in solvents (DMAc and NMP) and stirred for 2hours at 70°C. PAMPS was then added and the blend was stirred for 24hours. The solution was sonicated for various periods of time ranging from 10 to 90minutes using an Ultrasonic bath. In untralsonic bath, the final temperature achieved for a given time of sonication depends on the tempearutre of the laboratory (summer or winter). It is therefoe genrally advisable that maximum temperature of the bath is determined and maintains (as the so - called equilibrium tempearture), when operating continuously under ambient conditions (Santos et al. 2009). During the sonication, equilibrium temperature was maitained by simply filling the bath with water that was preheated to sample temperature.

Membrane Characterization

Water uptake and morphological tests were performed. Water uptake was carried out by first vacuum drying all membrane samples at 100°C until constant weights were obtained. Dried samples were then immersed in deionized water for 24h at room temperature (25°C). Samples were taken out of the water after 24h, wipped to remove surface water and then weighed immediately. Water uptake was calculated using Eq. 1.

$$water-uptake=\frac{W_{wet}-W_{dry}}{W_{dry}} \qquad \text{Eq. 1}$$

Surface morphology of the membrane was done using scanning electron microscopy (SEM). Thermogravimetric analysis (TGA) was done from room temperature to 350°C using TA Q600 analyzer at heating rate of 20°C/min under nitrogen atmosphere with flow rate of 20mL/min using ~ 8 -10mg membranes samples.

Results and Discussion

Membrane Morphology by Scanning Electron Microspcopy (SEM)

Results of surface morphological analysis performed on membrane were shown in Figure 3 and 4. The overall pictures of this figure depicts a dense membrane with some surface defects observed for most of the sonicated samples. For example, Figure 3A showed some kind of layered arrangement with no visible pores at the surface. Also, the PAMPS seems to be well distributed throughout the SPEEK. The micrograph in Figure 3B can be observed to be rough and possesses some uneven ramdom distribution of PAMPS agglomerates indicating that the distribution of PAMPS have been disturbed by the 10minutes sonication. An increase in sonication time to 20 minutes was observed to improve the distribution of PAMPS. This was however accompanied by some cracks believed to be due to the slight increase in the viscousity of the blend solution. The vicousity increase was brought about by the excape of some fraction of the solvent during sonication. As the sonication time was increased, PAMPS was observed to become more evenly distributed and the cracks become more visible and larger in size (Figure 3D to F). SEM images of samples at sonication time of 10 and 30 minutes were reproduced in higher magnification of 10μm (Figure 4) in order to further revealed the effect of sonication on the morphology of the somposites. Similar morphological behaviour has been reported in literature, though for inorganic/organic clusters of nanoclay/epoxy composites membranes (Lam et al. 2005).

Water Uptake

Water uptake have been reported to be affected by the degree of sulfonation, membrane pretreatment, hydration, ambient relative humidity and water temperature (Xing et al. 2004; Wu et al. 2006). Figure 5 showed the effect of sonication time and solvent type on water uptake that was measured at room temperature (25°C). For both solvents, water uptake is same at zero sonication time. There was however a further decrease in the water uptake for NMP at 20 minutes sonication time before it beging to increase. Thus, the best sonication time for both solvent lied between 10 and 20 minutes with respect to water uptake. Interestingly, this result correspond to that reported by Lam (2005) on the effect of ultrasonication on microhardness of composites.

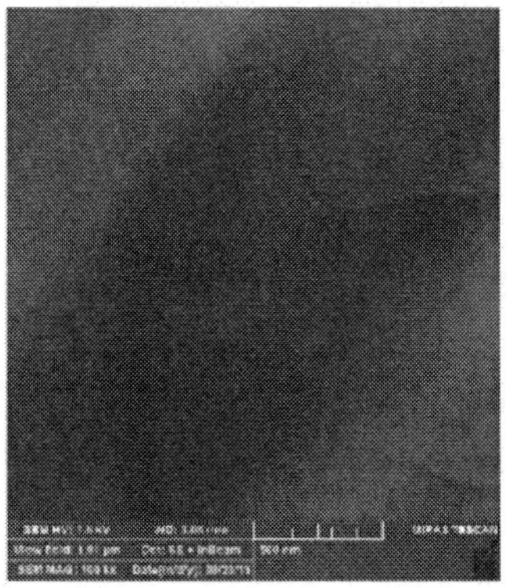

A) No sonication

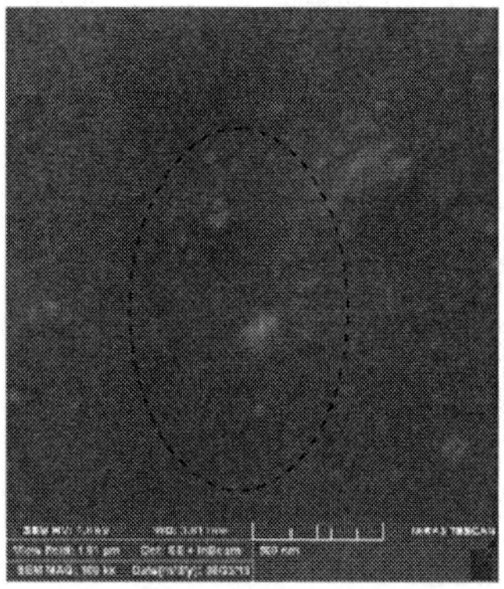

B) 10 minutes sonication

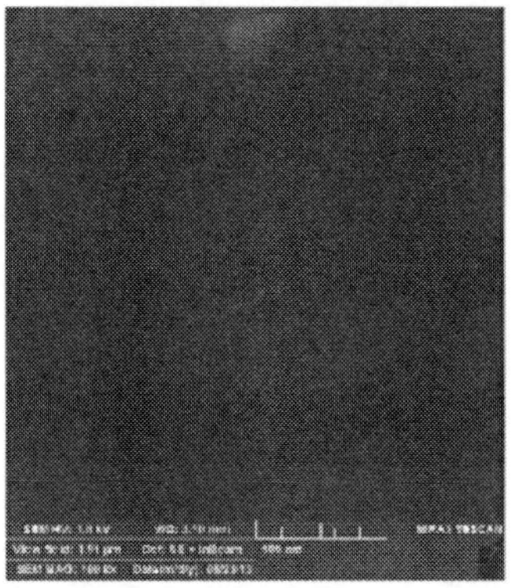

C) 20 minutes sonication

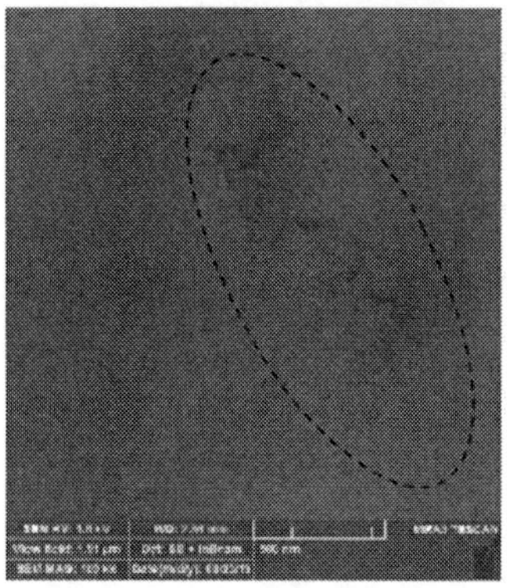

D) 30 minutes sonication

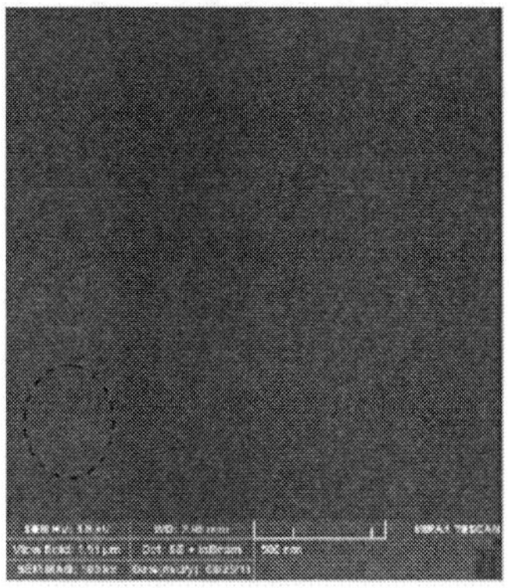

E) 60 minutes sonication

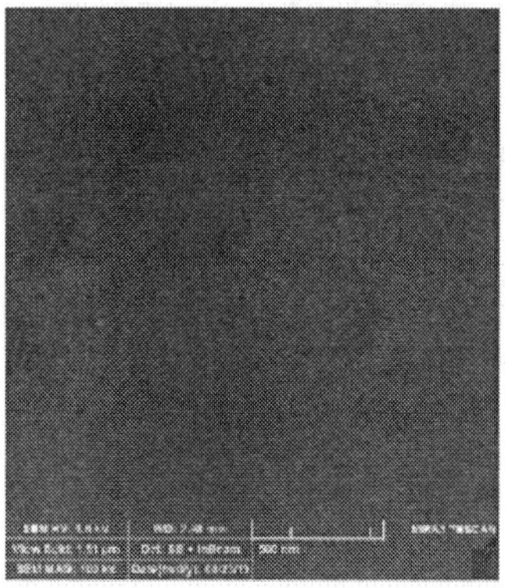

F) 90 minutes sonication

Figure 3 SEM micrograph of SPEEK-PAMS composite with various times of sonication with magnification of 500nm

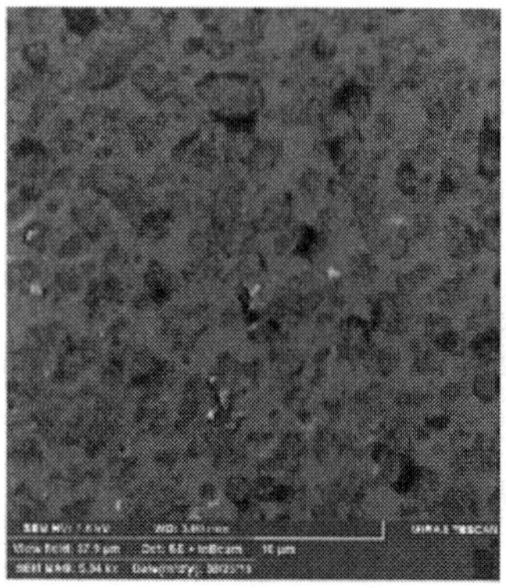

A) 10 minutes sonication

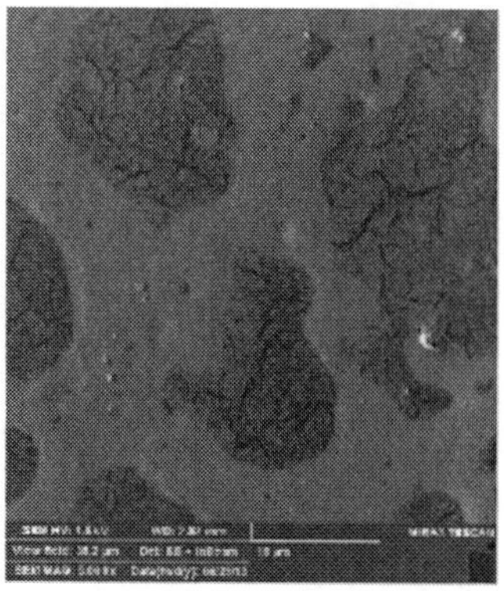

B) 30 minutes sonication

Figure 4 SEM micrograph of SPEEK-PAMS composite with various times of sonication with magnification of 10μm

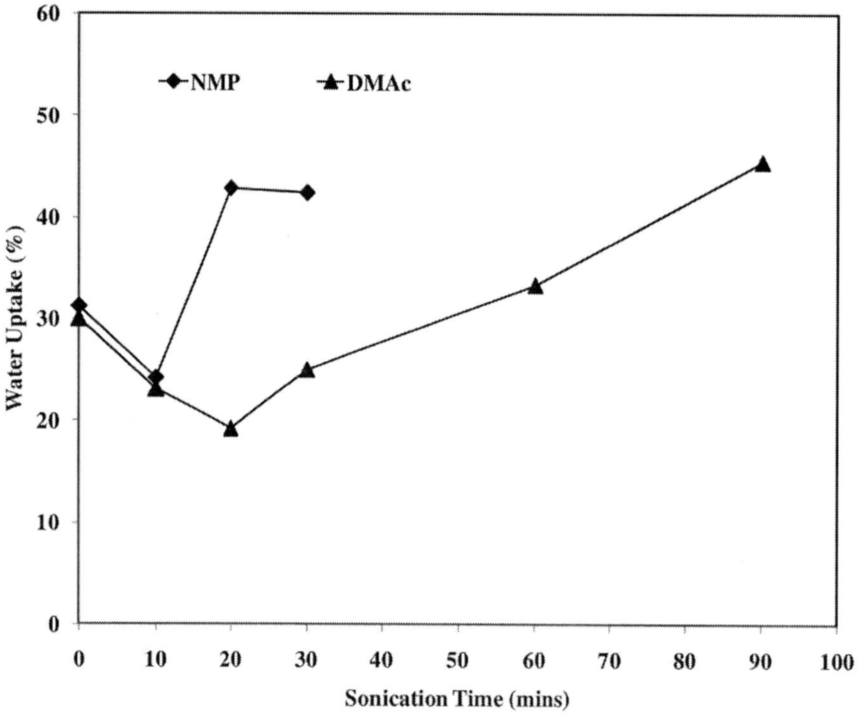

Figure 5 Water uptake of SPEEK-PAMS (3wt%) composite with variuos sonication times

Thermogravimetric analysis (TGA)

The difference in membrane performance resulting from the use different casting solvents have been attributed to the presence of residual solvents in the membrane samples after evaporation (Joly et al.

1999). TGA was thus used to quantify the residual solvent and evaluate its effect on the membrane performance as has been previously demonstrated in lieterature (Recio et al. 2007; Iqbal et al. 2008). Figure 6 shows the TGA curve for samples prepared from NMP and DMAc with different post vacuum heat treatment conditions. From this figure, weight loss can be attributed to evaporation of variety of substances, including solvents, from the samples (Rich et al. 1997). Table 2 contains the solvent and its boiling points, and weigth loss at the boiling point range that were extracted from TGA results. For sample heat treated for 24hrs, the overall percentage weight loss up to 204°C was about 8wt% of membrane sample for NMP while it is about 6.3wt% for DMAc. These will definitely include compnenents such as water, solvents and other unidentified substances. To quantify residual solvents, weight loss was estimated between 192 -214°C for SPEEK-PAMPS-NMP and between 154 and 176 for SPEEK-PAMPS-DMAc. The results obtained showed that the amount of the residual solvent in NMP is larger than that of DMAc by 1.25wt% (Table 1). To further investigate if this difference is responsible for higher value of water uptake for the NMP sample, membrane samples made from both solvents were further heat treated under vacuum for 72hrs. Residual solvent after this treatment was slightly higher in DMAc than NMP. Despite this, the value of water uptake of the NMP membrane is still higher than that of DMAc by approximately the same value as before the heat treatment (1.71wt%).

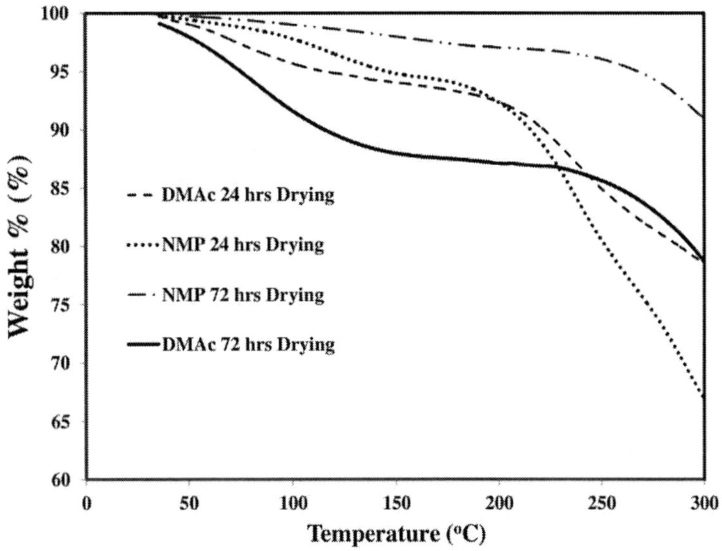

Figure 6 TGA of SPEEK-PAMPS-NMP and SPEEK-PAMPS-DMAc

Table 1 Residual solvent in SPEEK-PAMPS composites membrane samples

Membrane Sample	Solvent	B. Pt. (°C)*	Residual Solvent (wt%)	Water Uptake
DMAc 24 hrs Drying	DMAc	164.5 – 166	0.57	30.00
NMP 24 hrs Drying	NMP	202 – 204	2.85	31.25
DMAc 72 hrs Drying	DMAc	164.5 – 166	0.367	21.21
NMP 72 hrs Drying	NMP	202 – 204	0.217	22.92

*Boiling Points of Solvents

On a general note, results from Figire 1 – 5 showed that the best perfomance can be obtained for variety of sonication time depending the property of preference. Also, performance of DMAc solvent was observed to be better than that of NMP on the basis of water uptake. For this reason, DMAc was selected as the casting solvent for membrane preparation on which further studies will be carried out. Water uptake of DMAc membrane, prepared without sonication, was comapred with those reported in previuos literature. Table 2 illustrates the comparison between SPEEK-PAMPS and other relevant membranes previuosly reported in literature. In addition, this table serve to provide insight into the possible potentials of the composites. From this table, it was observed that the water uptake for SPEEK-PAMPS is within the range of values of other membranes on the table. Its water uptake is less than that of Nafion 112 which is a commercial membrane. The differences observed between the water uptake values of SPEEK – PAMPS and other membranes from the table can be explained based on three important factors - SO_3H group, the degree of sulfonation and experimental temperature (Gil et al. 2004; Shan et al. 2006; Wu et al. 2006). Water uptake of SPEEK – PAMPS is higher due to the presence of SO_3H group in the PAMPS molecules (Figure 2). SO_3H group have been reported to enhance membrane acidity and hydrophilic properties (Shan et al. 2006). Infact, the higher the SO_3H group the larger the water uptake. The degree of sulfonation has been reported to affect the water uptake of membranes. The higher the degree of sulfonation, the larger the value of water uptake (Gil et al. 2004). Thus, water uptake of SPEEK – PAMPS is lower than SPEEK -3 due their different degree of sulfonation, 53.3 and 72% respectively and the experimental temperature 24 and 60°C, respectively. The experimental temperature of SPEEK – 2 is close to that of the composite. However, the results cannot be compared in this case since the DS of SPEEK – 2 was not known. Above explanation is also true for the values of water uptake for SPEEK – 1.

Table 2 Comparison of water uptake values obtained with those from Literature (Xing et al. 2005)

Membrane	Water Uptake (wt%)	Experiment Temperature (°C)	**D S (%)	Reference
Nafion 112	22.85	N/A	-	(Xing et al. 2005)
SPEEK – 1*	16.00	30	45	(Ren et al. 2005)
SPEEK – 2*	44.90	25	N/A	(Zhong et al. 2008)
SPEEK – 3*	85.00	60	72	(Wu et al. 2006)
SPEEK– 4*	15.00	N/A	N/A	(Handayani and Dewi 2011)
SPEEK/PAMPS (3wt%)	21.21			This work

*The numbering have been included for explanation purposes.
**DS means Degree of Sulfonation

Conclusion and Recommendations

Blend of SPEEK and PAMPS was prepared by solvent casting method and its performance was evaluated with respect to its morphology and water uptake. Effect of sonication time (10 – 90minutes) and solvent types (DMAc and NMP) were also investigated. Performance of membrane prepared using DMAc was observed to be better than that of NMP. There was no correlation between the sonication time and membrane properties, but it was observed that the sonication deteriorate the membrane morphology. Membrane prepared without sonication shows better surface morphology, whereas the best water uptake was observed between 10 and 20 minutes. Thus, other properties need to be investigated in order decide the best sonication time. Therefore, further investigation such as thermal stability, methanol permeability, ion exchange capacity and gas (O_2 and H_2) permeability of the membrane are recommended for a complete evaluation of the membrane performance for direct methanol fuel cell (DMFC) application. All these investigations are in progress in our laboratory.

Acknowledgement

Authors would like to thank the National Science, Technology and Innovation Program (NSTIP) of KACST for funding this research under project number 10-ENE1374-04. Also, authors extend their appreciation to Center of Petroleum & Minearls, Center of Excellence in Renewable Energy, Mr. Dan Jerry D.Cortes of the Nanotechnology Laboratory, Chemical Engineering Department, King Fahd University of Petroeleum & Minerals, Saudi Arabis for their supports.

REFERENCE

Florjańczyk, Z., et al. (2003). "Proton conducting electrolytes based on poly(2-acrylamido-2-methyl-1-propanesulfonic acid)." Electrochimica Acta **48**(14–16): 2201-2206.

Gil, M., et al. (2004). "Direct synthesis of sulfonated aromatic poly(ether ether ketone) proton exchange membranes for fuel cell applications." Journal of Membrane Science **234**(1–2): 75-81.

Handayani, S. and E. L. Dewi (2011). "Influence of Silica/Sulfonated Polyether-Ether Ketone as Polymer Electrolyte Membrane for Hydrogen Fueled Proton Exchange Membrane Fuel Cells." Internat. J. of Sci. and Eng. **2**(2): 26-30.

Iqbal, M., et al. (2008). "Solvent effect on morphology and CO2/CH4 separation performance of asymmetric polycarbonate membranes." J. Membr. Sci. **318**(1–2): 167-175.

Joly, C., et al. (1999). "Residual solvent effect on the permeation properties of fluorinated polyimide films." Separation and Purification Technology **16**(1): 47-54.

Lam, C.-k., et al. (2005). "Effect of ultrasound sonication in nanoclay clusters of nanoclay/epoxy composites." Materials Letters **59**(11): 1369-1372.

Mohr, J. M., et al. (1991). "Gas transport properties of a series of poly(ether ketone) polymers." Polymer **32, 13**: 2387-2394

Recio, R., et al. (2007). "Gas separation of 6FDA–6FpDA membranes: Effect of the solvent on polymer surfaces and permselectivity." J. Membr. Sci. **293**(1–2): 22-28.

Ren, S., et al. (2005). "Surface modification of sulfonated poly(ether ether ketone) membranes using Nafion solution for direct methanol fuel cells." Journal of Membrane Science **247**(1–2): 59-63.

Rich, D. C., et al. (1997). "Effect of cure conditions on Probimide 32 polyamide-imide." Journal of Applied Polymer Science **63**(9): 1113-1126.

Santos, H. M., et al. (2009). The Power of Ultrasound. Ultrasound in Chemistry: Analytical Applications. J.-L. Capelo-Martinez. Weinheim, Wiley-VCH.

Shan, J., et al. (2006). "Sulfonated polyether ether ketone (PEEK-WC)/phosphotungstic acid composite: Preparation and characterization of the fuel cell membranes." Pure Appl. Chem **78**(9): 1781 - 1791.

Swier, S., et al. (2005). "Sulfonated poly(ether ketone ketone) ionomers as proton exchange membranes." Polymer Engineering & Science **45**(8): 1081-1091.

Wu, H.-L., et al. (2006). "Preparation and characterization of poly(ether sulfone)/sulfonated poly(ether ether ketone) blend membranes." European Polymer Journal **42**(7): 1688-1695.

Xing, D. M., et al. (2005). "Characterization of Sulfonated Poly(Ether Ether Ketone)/Polytetrafluoroethylene Composite Membranes for Fuel Cell Applications." Fuel Cells **5**(3): 406-411.

Xing, P., et al. (2004). "Synthesis and characterization of sulfonated poly(ether ether ketone) for proton exchange membranes." Journal of Membrane Science **229**(1–2): 95-106.

Ye, Y.-S., et al. (2012). "Water Soluble Polymers as Proton Exchange Membranes for Fuel Cells." Polymers **4**(2): 913-963.

Zhong, S., et al. (2008). "Modification of sulfonated poly(ether ether ketone) proton exchange membrane for reducing methanol crossover." Journal of Power Sources **180**(1): 23-28.

Fixed Plant Facility Siting – A Mega Project Lifecycle Framework

M.F. Haider
mech_haider@yahoo.com
mhaider@suncor.com

Prepared for Presentation at
American Institute of Chemical Engineers
2014 Spring Meeting
10th Global Congress on Process Safety
New Orleans, LA
March 30 – April 2, 2014

Fixed Plant Facility Siting – A Mega Project Lifecycle Framework

M.F. Haider
mhaider@suncor.com
mech_haider@yahoo.com

Keywords: Facility Siting, Project Management, Mega Project, Consequence Analysis (CA), Quantitative Risk Assessment (QRA), Adaptive, Change driven, Agile methodology

Abstract

Facility Siting based on tools such as Consequence Analysis (CA) and Quantitative Risk Assessment (QRA) is often times successfully applied for Brownfield operating unit projects, projects with relatively smaller footprints and for projects lasting relatively short time-periods. Challenge remains to apply and realize the full potential and value of Facility Siting for Mega-Projects. This is mostly due to project complexity and length – this paper will introduce a Management Framework to successfully apply CA and QRA as part of Facility Siting for Mega-Projects. The paper will also outline key strategies to ensure successful (i.e. value adding) implementation of facility siting, suggest a list of activities, associated sequence and pitfalls to avoid. Project Management team (e.g. Project Sponsor, Project Engineers, Managers) of both owner/operator and contractor/vendor may find this paper more useful as it provides a broad overview (i.e. when to initiate, how to plan, execute and close such assignment to maximize value) including real life examples of facility siting process over the project lifecycle.

1. Introduction

The driving force for process safety has been primarily based on catastrophic events [1]. With an increasing number of incidents, the process industry and governments started to take measures such as Facility Siting initiatives with hope to minimize loss of life and property. In the 1970s and 1980s, some of the world's most tragic industrial accidents took place. The Flixborough explosion in 1974 was reinforced by that of the Seveso tragedy in 1976 [2]. The Macondo blowout in the Gulf of Mexico on April 20, 2010, killed 11 employees and led to an uncontrolled oil spill lasting 87 days. This blowout was the most significant offshore incident in the US to-date. Finally, the Fukushima Daiichi nuclear plant incident in Japan in March 2011 drew the attention of the global process and power industries and encouraged them to incorporate natural disaster risks in a hazard analysis study [3]. As part of inherently safer design, many companies now-a-days apply CA and QRA as part of facility siting for their new projects. Due to the scarcity of information at the beginning, the studies in many cases start later in the project lifecycle for them to be effective and value adding. This paper will focus more on how to manage such initiatives for new Mega-Projects over the technical intricacies of the CA/QRA processes. The intention of the paper is to enhance the Body of Knowledge publicly available for management team in as much non technical languaga as possible to make their greenfiled Mega-Project initatives a success i.e. safer for people, environment and property.

So, what is different for industrial Mega-Projects?

- Industrial Megaprojects, as the name implies, are very large. To provide a simple and simply applied definition, we are defining a megaproject as any project with a total capital cost of more than $1 billion (U.S. dollars) as measured on January 1, 2003 [4].
- Most often engineering and construction activities run in parallel and not in sequence
- Multiple sub-projects under the 'Mega-Project umbrella' run in parallel
- Complex – design is almost 'never' complete in early stages – goes through many iterations
- Progressive elaboration is evident
- Scarcity of critical workforce with mega-project experience
- Scarcity of information – requires project team to make assumptions for CA/QRA
- Timeline: 2-5 years

2. Key Strategies

2.1 Facility Siting Charter

It is not only important or good to do but critical that the stakeholders need to identify, understand and agree on the purpose and value to conduct facility siting for their projects. The Author's personal experience over the years has been in order to maximize value of Facility Siting the project team has to start 'early' to develop project Charter and start with the end in mind. As can be seen from the following generalized Figure 1 that as the company starts to appraise opportunities, it is

Figure 1: Stages and Milestones to apply CA/QRA in a Mega-Project

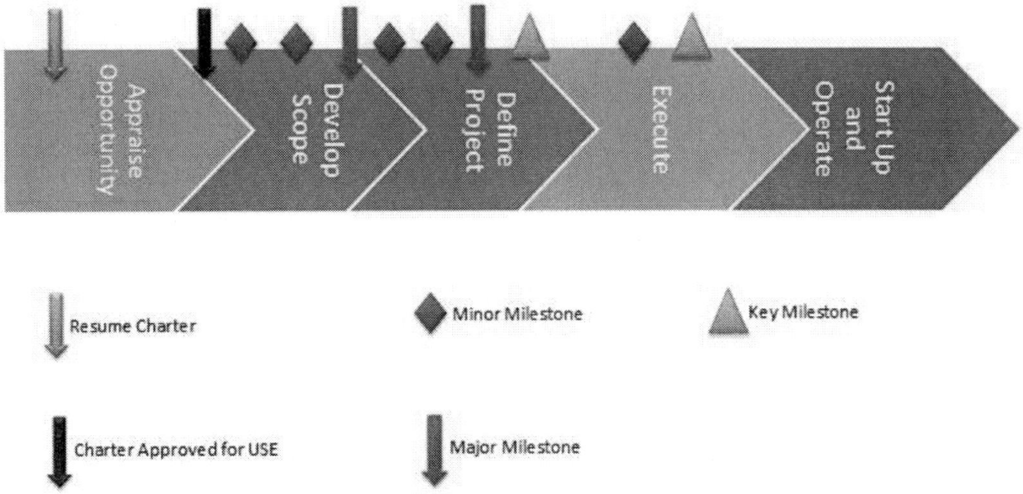

prudent to resume developing the Facility Siting Charter by bringing in the stakeholders together. Figure 1 shows a generalized project lifecycle for a typical Petro-Chemical Mega-Project. The

Figure also shows that the Key Deliverable is developed through a successive series of repeated cycles over the project life-cycle. The Milestones will be defined later in the paper. Part of starting with the end in mind is to clearly outlining ultimate goal/s, e.g. vision is to design a safer facility and utilize Mitigated QRA Report to formally document and communicate that facility is within the Risk Threshold or justification for accepting risk profile as is by Senior Management. The Vision in the Charter also provides opportunity to strike a balance between business goals and climbing the Risk Mountain – at each stage of increasing safety (decreasing risk). As part of striking a balance, the Project Engineer has to Identify options up-front if and/or when it becomes apparent that aggregate Residual Risk will not be within the Risk Threshold or acceptable so that team can put more effort in developing justification for either accepting or rejecting the risk profile by Management

Figure 2: Strike a balance between Business goal and Climbing the *expensive* Risk Mountain

2.2 *Management Strategy*

In order to successfully manage Facility Siting, CA and QRA for a Mega-Project, the stakeholders have to align on a management strategy. There are two primary management approaches for the purpose of managing such work:

1. Predictive or Plan driven
2. Adaptive or Change driven

The Author's past experience has been if the organization undertook very 'similar' projects in the near past (<5 years), critical experienced workforce is retained, Lessons Learned is readily available, and most importantly, the past project is still considered to be a success, it is advisable to use Plan driven approach. On the other hand, if the organization is introducing new technology, has limited past experience with similar project, involves relatively high number of stakeholders, capital cost > $1B, project life spans >2 years, it is recommended to utilize Change driven approach to maximize value. It is to be noted that it is not essential to meet all the suggested criteria mentioned above to pick one management strategy over the other.

2.2.1 Benefits of Adaptive or Change Driven Approach

In Adaptive/Change Driven Planning, project team intentionally repeats one or more project activities as the project team's understanding of the product increases ....Also known as ...agile methods .....intended to respond to high levels of change and ongoing stakeholder involvement [5]. The heart of Facility Siting involves making timely but informed decision. Some of the characteristics of this framework:

1. To add small incremental improvements that will deliver value to stakeholders
2. Milestones develop the product (e.g. Mitigated QRA Report) through a series of repeated cycles, while increments successively add to the functionality of the product
3. For each Milestone, a deliverable (read 'model') or set of deliverables will be completed. Future iterations may enhance those deliverables or create new ones. Each iteration incrementally builds the deliverables until the exit criteria for the phase are met, allowing the project team to incorporate feedback.

The biggest challenge with Mega-Project is scarcity of key information needed as inputs to develop scenarios and models. Some of the critical benefits, a project may gain by applying Change Driven approach:

1. The approach acts as Leading Indicator on Process Safety for Plant Layout - allows for multiple opportunities to make critical decisions on plant layout over the project life-cycle capitalizing on what is known at a given time. This lets the project team to use "what they know", to ensure "what they don't know" falls into the right place
2. Opportunity to make decisions over the project lifecycle not only works to save life but also helps with striking a balance between business needs and climbing the risk mountain
3. Aides design-development with progressive elaboration while producing final CA/QRA deliverable for the project

2.3 Execution Strategy

The main idea of Change Driven process starts with decomposing the known preliminary layout and process information into a set of requirements, work to be performed and their priority sequence, sometimes referred to as a *Product Backlog*.

It is advisable that the owner/operator forms a multi-use agreement (contract) with the chosen vendor. This way the plant areas under the Mega-Project umbrella can get their specific work done while the CA/QRA Project Team provides overall oversight and guidance.

At the beginning of an iteration, the team will work to determine how many of the highest priority items on the backlog list can be delivered within the next iteration. The time-length of each iteration should be less than 4 weeks. As shown in Figure 1, during "Develop Scope" and "Define Project", the length of each iteration could be even less than 2 weeks. This iterations (otherwise known as "Minor Milestone" as shown on Figure 1) will develop scenarios, assumptions and preliminary CA/QRA models. Assumptions are required due to lack of available quality information in the early stages of a project.

At the end of each iteration, the Minor Milestone/deliverable should be ready for review by the customers/stakeholders. This does not mean that the customer is required to accept delivery, just that the product should not include unfinished and/or incomplete features.

The sponsor and customer representatives should be continuously engaged with the other project stakeholders to provide feedback to Contractor on deliverables as they are created and to ensure that the *Product Backlog* reflects Sponsor/stakeholders' current needs. Below, you will find two examples using two different management strategies. For confidentiality purpose, the specific name/location is kept generic and few items were modified to suit.

2.3.1 Project X Facility Siting Using Change Driven Approach

A CA/QRA team was given the assignment to support design development during "Develop Scope" phase (Figure 1). The Sponsors already selected the site for the project. But the fixed plants were yet to be located on the site. The Vision for the initiative was for the QRA result to come under the organization Risk Threshold. Multiple EPC's were given to engineer different plant areas. The CA/QRA team selected a vendor to support them with defining the *Product Backlog* based on the preliminary Process Design Basis, intended product the facility is expected to sell, the feedstock necessary etc. The CA/QRA team had prioritized the Product Backlog based on hazard level of feedstock, preliminary PHA schedule and project schedule. This also involved obtaining feedback from different Process Engineering and Operations teams to collect necessary inputs to put together the assumptions and scenarios to be modeled within the 1st iteration.

Out of 25+ scenarios, only 7 perceived to be the most hazardous were modeled for CA that gave ample input (e.g. explosion, toxic and fire contours) for the CA/QRA team to work with the stakeholders to recommend location for specific process units on the given site. Note that, CA/QRA team integrated the potential consequence to understand local incident and global consequence events. At this time QRA was not performed. This was used as input for minor modification of preliminary plant layout. It is to be noted that the site plan was still preliminary after the first iteration but assisted different EPC accountable for their specific area to recognize where to locate the specific plants i.e. finalized key plans for the overall site plan.

With further design-development, more refined information became available from each EPC. This new information led to 130+ scenarios and was used for further CA/QRA iterations that created either new model and/or embellished the past ones. Each EPC was accountable for facility siting within their boundary while the CA/QRA team provided technical oversight and integration support for the overall project. The Sponsor/stakeholders were engaged on a regular basis to ensure that the Product Backlog reflects future modeling requirements to properly site facilities e.g. Fire Water Pump house, Tank Farms, site Medical Centre etc. Early engagement of CA/QRA team and higher management willingness to incorporate their feedback with the progressive elaboration of the project led to QRA result meeting the organization risk threshold and ultimately safer facility. Changes suggested based on CA/QRA work in earlier stage (i.e. Develop Scope, Define Project) were easier to incorporate into the design.

2.3.2 Project Y Facility Siting Using Plan Driven Approach

The sponsors for project Y selected the site for the project and decided to hold on to not perform CA/QRA work with the progressive elaboration of the project activities. This was mostly decided based on the fact that not all information was available to properly conduct the 'studies'. Fundamentally, the sponsors looked CA/QRA as deliverables requiring a more 'complete' plan. Once the Develop Scope (as referred in Figure 1) is almost complete, the sponsors decided to engage CA/QRA team and a vendor. By this time, the project had significantly high number of stakeholders. Many of the stakeholders were not very familiar with CA/QRA and the value it could bring to them. There was not much time to schedule and train the stakeholders as well to bring them up to speed. As a result, the stakeholders were unable to align on a Vision. The Vendor was asked to provide explosion, toxic and fire contour maps for specific scenarios and locations that were almost 'fixed' on the site plan at that point in time. The resultant contour maps showed significant consequence on nearby facility with over 700+ people. It was also noted that the plant area consuming the most hazardous substance would impact the nearby facilities significantly from consequence perspective. It should be noted that the result was only based on a selected few scenarios. By the time the project reached the "Execution" phase, more scenarios were added as part of the scope. But it was already too late to incorporate most of the recommended changes resulting from CA/QRA work. The project went through several area by area risk reduction workshops only to accept (by the company executives per organization risk matrix) the aggregate residual risk level which was also found to be not within the organization risk threshold.

2.3.3 Importance of Starting Early

Figure 3: Impact of variable based on project time [5]

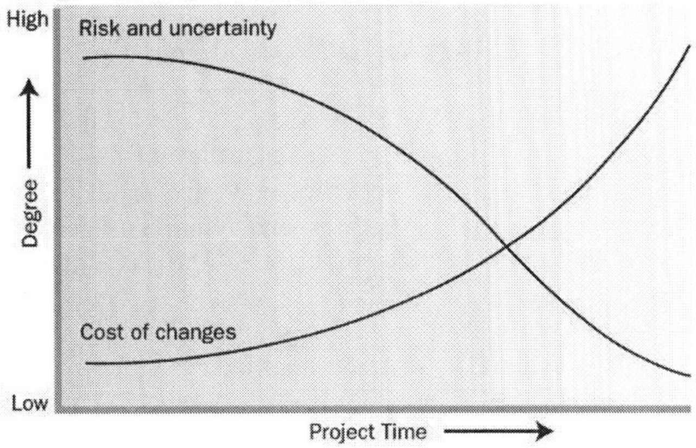

The above two examples highlight the importance of starting CA/QRA work early enough to realize and maximize the value CA/QRA brings to a Mega-Project. The high-level view provides common frame of reference for comparing projects for CA/QRA initiative, even if the projects were dissimilar. Change driven methodology allows for applying the known information at early stage in a way that supports laying out plant facilities to reduce risk and thus maximize potential value. While these characteristics remain present to some extent in almost all project life cycles, they are not always present to the same degree [5]. Change Driven framework, in particular, keeps stakeholder influences higher and the costs of changes lower especially in the early stages of the life cycle than in predictive/plan life cycles. The Figure 3 highlights a well-known fact of project management that even though risk and uncertainty remains high in the early stages of a project, cost of incorporating changes remains much lower. This translates into a key opportunity for project stakeholders to align on laying out the key areas on a site plan intelligently to reduce risk and maximize value by applying tools such as CA and QRA.

2.4 Key Decision Points

Facility Siting, from a management point of view, is all about bringing the stakeholders closer and find the 'right' spot for the facilities. Finding the 'right' spots requires making timely decisions. The challenge has always been to make timely decisions as the stakeholders more often find themselves forced to make those decisions based on information that are frequently of poor quality and at times uncertain, especially in the early stages of a project. At times, due to the scarcity of information, stakeholders rely on their past experience to make those decisions on plant layout. As a result, depending on the individual experience, two different teams for a project area may come up with two very different plant layout. CA/QRA work, if started early enough, will also depend on input that most often requires embellishment through assumptions to create needed scenarios and models. Even with this limitation, CA/QRA provides a key 'input' in allocating real estate for key facilities on a plant site at early stage of a project lifecycle. With each iteration and deliverable 'Change driven' methodology allows for multiple opportunities to make critical decisions on plant layout over the project life-cycle. It allows the stakeholders to decide when to 'sharpen the pencil' (e.g. PHAST to Computational Fluid Dynamics model), relocate facility or accept the consequence and/or risk, 'pull the plug' on further modeling/refinement and start developing justification, if required, to recommend management to accept/reject the aggregate risk profile.

2.5 Alignment with Process Hazard Analysis (PHA) Planning

CA/QRA results are a useful input to PHA (e.g. HAZOP, LOPA) activities. This is particularly true as projects move forward after establishing the plant layout on a high level and starts to conduct "Define Project" (Figure 1). PHA activities may also result in recommendations (i.e. new scenarios) that goes into the *Product Backlog* and requires project team to develop new models or refine old ones. Mega-projects have multiple sub-projects running in parallel most of the time, having an Integrated Schedule for all the reviews is a good strategy so that the right stakeholders are booked for the sessions. The integrated PHA schedule also supports prioritization of *Product Backlog* for CA/QRA activities over the lifecycle of the project.

2.6 Integration Strategy

It is critical for a Mega-Project involving multiple plant areas/silos to understand the BIG picture so that site risk can be dealt with on an aggregate basis. There is strong possibility of one area to impact other plants in the vicinity. It is recommended to review integration aspect and aggregated results on a regular basis. The following Figure 4 shows that the origin of an incident may reside in one particular plant area, while it may impact other areas in the vicinity.

Figure 4: Consequence of an incident in one area impacting multiple areas

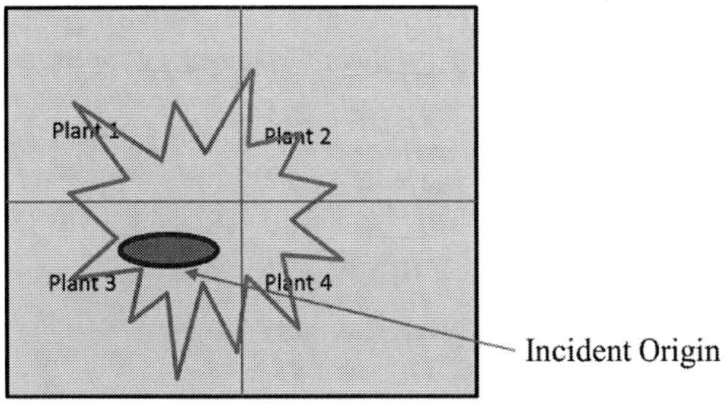

Recommended Strategy:

- Setup Strong Center - Task Force Single Point of Contact (SPOC)
- Collaborate – SPOC will promote Collaboration among Plants/Areas

3. List of Activities (suggested)

3.1 Overall Sequence

The CA/QRA work typically should start with the approved Charter as shown on Figure 1. A project Task Force i.e. SPOC is formed comprising of Project Engineer, Operations rep, Process rep and CA/QRA Technical Expert. The SPOC is ultimately accountable to produce the CA/QRA deliverable studies but also acts to support the design development and site plan finalization efforts in the early stages of the project.

The team collects 'input' such as Preliminary site Layout and design information, works with area process and operations team to identify scenarios, develop assumptions for known-unknowns and unknown-unknowns, develop modeling 'Product backlog', prioritize backlog based on PHA schedule and produce modeling output (i.e. deliverable) for stakeholder use. The stakeholder usually has the option of accepting or rejecting consequence. If the result of an iteration is acceptable within area causing the incident – communicate to outside stakeholders to ensure there is no cross boundary impact. If the result of the iteration is NOT acceptable to outside of the area – use the result as input to PHA for the areas to mitigate. Collaboration recommended between origin area and impacted area/s to identify appropriate mitigations. There is also a possibility that the consequence resulting from the iteration is not acceptable to the area causing the incident, in that case the area will initiate a PHA to mitigate the consequence. As part of developing mitigations, consider the following:

- Develop QRA if appropriate
- Perform Building Assessments for Critical Buildings
- Apply Cost-Benefit analysis to optimize risks/opportunities

- Determine Safe Haven locations and their characteristics

If Mitigation is not possible from Business perspective – perform As Low As Reasonably Practicable (ALARP) justification for Senior Management to formally accept the posed risk

- To apply mitigations on a site wide basis, review Milestone deliverables by taking on a BIG picture perspective
- Future iterations may enhance those deliverables or create new ones as new information is available. Each iteration incrementally builds the deliverables to produce the Mitigated CA/QRA report to support Facility Siting for the stage

The overall effort is expected to produce a maximum of two single reports for the overall site:

1. Mitigated version of QRA Report
2. and Consolidated CA Report

It is recommended to develop as-built, pre-start up CA/QRA reports prior to start-up. Table 1 and 2 outlines typical activities performed by SPOC and plant areas respectively once key process areas are established/finalized on Site Plan. Please note that the activities are project dependent and may warrant modifications to suit specific situation.

Table 1. SPOC Sequence of Activities to conduct Facility Siting

Initiative	Key Steps
Setup Facility Siting Team	Sponsor to support to setup a team (i.e. Project Engineer, Technical Expert, Operation, Process representative) to act as SPOC
Integration	Provide functional (technical and management) guidance to Areas
	Assess cross boundary impact/opportunity
	Communicate incremental outputs after each Area PHA/Risk Reduction
IFA versions (prior to "Define Project" stage)	Consolidated CA Report
	Mitigated and Integrated QRA Report
IFD CA/QRA (prior to "Execution" stage)	Re-validate IFA CA/QRA post PHA and vendor PHA
As-built versions	Post-oil version

3.2 Typical Area Activities

Table 2. Area (Typical) Sequence of Activities to conduct Facility Siting

Initiative	Key Steps
Plant 1 (Typical)	Perform Modeling and Calculations
	Conduct Risk Reduction Workshop (if appropriate) – Apply mitigations
	Incremental output for use based on reduced risk level
	Assess Buildings/facilities
	Identify Safe Haven/shelter Locations (if required)
	Estimate Area Project Scope impact
	Perform area scope updates/changes e.g. passive fire protection requirement
	Carry out QRA (if appropriate)
	Update Product Backlog
	Develop/update contract to support area specific activities

4. Pitfalls to Avoid

The following observations were noted by the Author during his time to manage Facility Siting scope:

1. Facility Siting does not replace good and common sense engineering practice e.g. locating a plant that uses high volume of hazardous material will significantly increase risks to other plants/areas (e.g. camp) within the vicinity
2. Evaluate effectiveness of 'Cost Reduction' philosophies as opposed to added risk e.g. different modularization techniques may decrease the plant footprint by reducing pipe lengths but makes the plant more compact and likely to increase severity of explosion consequence
3. Avoid developing models with hope to realize risk reduction
4. Sharpening the pencil for models most likely may not reduce risk
5. Avoid jumping into CFD or other similar methods without evaluating cost-benefit for undertaking such effort
6. Utilization of QRA results is much more controversial than the methodology. Watson [6] has suggested that QRA should be considered as an argument, rather than a declaration of truth. In this view – focus on accuracy over precision
7. CA/QRA Task Force should be part of Management of Change (MOC) process
8. Evaluate carefully before locating facilities e.g. tank farms close to main entrance/egress as part of Facility Siting

9. Keep the focus on VISION for the overall effort to avoid scope creep
10. Note the difference between Facility Siting as a Deliverable and tool for design-development activities

5. Conclusion

This paper presented Adaptive/Change driven framework to manage initiatives like facility siting/CA/QRA as part of Mega-Projects. Change driven framework does not mean "lack of planning or no planning". As demonstrated in the paper, this methodology starts with a vision and then develops the ultimate product both iteratively and incrementally when it is possible to define small incremental improvements that will deliver value to stakeholders. This allows making key decisions on laying out the facilities in early stages of a project rather than waiting for design to develop to a significant extent. Often times, project team wait too long for design to mature enough before applying CA/QRA to validate consequence impact and/or risk level. Most often, especially for Mega Projects, this leads to delays in executing the CA/QRA work properly at the right time to influence site layout. This delay in execution leads to scenarios when scope changes cannot be incorporated anymore. This translates into value erosion as project was unable to reap the full benefit of Facility Siting. The management framework recommended in this paper i.e. Change driven/adaptive framework allows for the stakeholders to start early in the project lifecycle to apply CA/QRA in laying out the plant facilities. The key is to start EARLY but START with the END in mind. Alignment among stakeholder on a VISION up-front is also important as this is expected to save life while striking a balance between business needs and climbing the Risk Mountain. Vision also allows avoiding scope creep while applying Adaptive/Change driven approach. It is highly recommended that stakeholders focus more on accuracy over precision. In the early stages of project lifecycle, quality information is scarce. Change driven/Adaptive framework gives you as the project stakeholder the opportunity to use "what you know" intelligently to ensure "what you don't know" falls into the right place.

6. References

[1] M. Mannan, A. Chowdhury and O. Reyes-Valdes, "A portrait of process safety: From its start to present day," College Station, 2012.

[2] M. Mannan, Lees' Loss Prevention in the Process Industries, Volumes 1–3 (3rd Edition), , 2005, Elsevier, 2005.

[3] K. McAndrews, "Consequences of Macondo: A Summary of Recently Proposed and Enacted Changes to US Offshore Drilling Safety and Environmental Regulation," in *Society of Petroleum Engineers, Americas E&P Health, Safety, Security and Environmental Conference*, 2012.

[4] E. W. Merrow, Industrial Megaprojects Concepts, Strategies and Practices for Success, New Jersey.: John Wiley & Sons, Inc., 2011.

[5] A Guide to the Project Management Body of Knowledge (PMBOK® Guide) – Fifth Edition, Newtown Square: Project Management Institute, Inc., 2013.

[6] S. Watson, "The Meaning of Probability in Probabilistic Risk Assessment," *Reliability Engineering and System Safety,* vol. 45, pp. 261-269, 1994.

348480 Global Energy and Transportation and Mobil Oil

Abstract:

Biofuels, Computational Methods, Engineering Economics, Polymers, Thermodynamics. Polymers-plastic, Thermodynamics- oil, Engineering Economics – the money it takes to finish the method of Global Engineering Research, Computational Methods – numerical input of analytical data, Biofuels – the transport fluids that utilized in more than one method from industrial usage to everyday living which enters the atmosphere. Exxon has completed a $600 million dollar deal for production of algae biofuel, amongst Mobil Oil has performed major research Global Energy and the effects of oil. Reading a most recent article on Mobil website.

348748 Characterization of Iron Phthalocyanine As the Active Material for Lithium Batteries

Abstract:

Results on the determination of the lithium-ion diffusion coefficient in the iron phthalocyanine (FePc) active material as a function of intercalation level, using a modified electrochemical impedance spectroscopy technique, will be presented. Also, developed thermodynamic functions such as the Gibbs free energy, entropy, and enthalpy changes for the lithium insertion reaction with FePc, derived from cell open-circuit voltage measurements as a function of intercalation level, are presented. Also presented are estimates of the maximum amount of lithium that can be intercalated into the FePc based on constant volume and temperature conditions.

349086 Design of a Free-Fall Reactor for Fast Pyrolysis of Waste Plastics

Abstract:

Thermal pyrolysis has earned significant reputation in the recovery of valuable chemicals, especially petrochemical based fuels from various materials, including biomass and plastics. Of many available technologies, *fast* pyrolysis was preferred in his study owing to its minimal heat and mass transfer limitations and better control on process conditions. The focus of this study is towards efficient conversion of waste plastics into gaseous products consisting primarily of alkanes, alkenes and aromatics. Lab scale experiments were performed using a micro reactor to determine favorable conditions, including reaction temperature and particle size that would result in maximum yield of ethylene. Using these results, a simple industrial size radiatively heated free-fall reactor was designed for the purpose of pyrolyzing waste plastics. The goal is to design a *fast* pyrolysis reactor that improves upon inefficiencies in traditional reactors while resulting in high yield of ethylene.

349732 Biosorptive Dehydration of Ethanol/Water Azeotropes Using Compound Starch-Based Adsorbent

Abstract:

Abstract: Recently, the interest in the production of bioethanol has greatly increased as a result of the increasing fossil fuel prices and associated environmental issues. Bioethanol is made from either grains or biomass and the broth contains 6 to 12 percent ethanol. In addition, water removal represents a serious problem as ethanol form azeotrope with water at atmospheric pressure. Distillation to water-free ethanol consumes 50 to 80 percent of energy used in the fermentation ethanol manufacturing process. Various techniques have been developed to break the azeotropic point, such as salting-out method, azeotropic distillation, extractive distillation, reactive distillation, adsorptive distillation and pervaporation. Among the methods, the adsorptive option is the most attractive, especially the biosorption, which has distinct advantages over conventional methods as it is environmentally friendly, highly selective, easily available, easy to operate, cost effective and reusable in repeated cycles in the treatment of aqueous organics.

In this work, a specially formulated compound starch-based adsorbent (CSA), which consists of corn, sweet potato and foaming agent, was developed. As an exploratory study, this research was focused on the investigation of CSA for biosorptive dehydration of aqueous ethanol and optimization of conditions for its maximum adsorption under various conditions of vapor feed flow rate, bed temperature and bed height through an orthogonal design method. Furthermore, to investigate the adsorption mechanism, dynamic saturated adsorbance of CSA was studied and characterization of the adsorbent before and after adsorption, as well as after regeneration, was performed.

In the first step, the net retention time and separation factor of ethanol and water were obtained using inverse gas chromatography (IGC), the results indicated that the net retention time of water on CSA is significantly longer than that of ethanol which attested the feasibility of this adsorptive separation purpose and low temperature could be propitious to the process which was consistent with the traditional theory of adsorption.

Then, through a batch adsorption experiment, the optimum adsorption condition was determined under vapor feed flow rate, bed temperature and bed height by orthogonal design method, which indicated that the optimization condition for ethanol was at vapor feed flow rate of 1.1g/min, bed temperature of 83 and bed height of 25cm. The packing density was 445kg/m^3. The residence time is determined by vapor feed flow rate and bed height. The residence time is more influenced than bed temperature. The longer the residence time, the more the output is. But it cannot be ignored that the adsorption selectivity may decreases. Low bed temperature is benefited to the adsorption process which agrees with the result of IGC. Five cycles experiments of adsorption-regeneration were carried out at the optimum adsorption and regeneration condition, the results showed that there is practically no change in the regeneration efficiency of CSA after the cyclic process adsorption–regeneration. It is due to the properly chosen regeneration conditions assuring complete desorption of the water adsorbed without changes in the CSA structure. This unchanged adsorption capacity after a number of cycles demonstrates reusability of the adsorbent.

To study the adsorption mechanism, experiment was conducted to study the dynamic saturated adsorbance of CSA. The water adsorption capacity of CSA has been found to be similar to some molecular sieves, which is about 0.15g/g. Besides, field emission scanning electron microscopy (FESEM) and mercury porosimetry (MP) were applied to investigate the change of surface morphology and microstructure of CSA before and after the adsorption, as well as after regeneration, as shown in Fig.1 and Fig.2. The FESEM images showed that compared with the particles before the adsorption, they became plump and the void spaces became smaller after adsorption, which indicated that the CSA was saturated with water. However, the CSA particles after regeneration became wizened like the nature particles, which demonstrated that almost all the water had been desorbed from CSA. The pore size distribution (PSD) indicated that compared with the PSD curve before adsorption, the peak at 6000nm disappeared and the peak at 17000nm became remarkably smaller after adsorption, which manifested that CSA is mainly composed of macropores which play a major role in the adsorption. However, the peaks at 6000 and 17000nm reappeared after regeneration, and the curve looked exactly the same with that before adsorption, which further demonstrated that CSA after regeneration can be totally regenerated and recovered to the original structure.

These results indicated that the starch-based adsorbent can be considered as an effective and low-cost bio-sorbent to dehydrate alcohol.

349781 Purification of 2-Amino-1-Phenylethanol Enantiomers By a New Technique Combining Distillation and Crystallization

Abstract:

Stripping crystallization (SC) is a new separation technique which combines distillation and crystallization. Basically, SC is operated at a triple-point condition, in which the liquid is simultaneously vaporized and crystallized due to the three-phase equilibrium. In essence, the process is continued until the liquid phase is completely eliminated and only the pure solid crystals remain in the feed. Thus, crystal washing is not required since only pure solid crystals remain in the feed and no impurities are adhered on the crystal surfaces at the end of SC. In the previous research, SC has been successfully applied to separate p-xylene from the mixed xylenes by Shiau et al. (2005, 2008).

Phenylethanolamine, also known as 2-amino-1-phenylethanol (2-APE), is an important basic skeleton in many adrenergic drugs and is often used as ingredients of cold medicines and anorectics. As the biological activity and pharmacokinetics of the enantiomers of a chiral drug can be different, the objective of this research is to apply SC to purify R-2-APE from the chiral 2-APE mixture. The experiments show that no crystallization occurs during SC without seeding due to subcooling of the liquid enantiomers. On the other hand, seeding can eliminate subcooling during SC, leading to the R-2-APE crystals gradually formed preferentially on the seeds. Thus, seeding with magnetic stirring can be successfully employed to purify R-2-APE from the 2-APE enantiomers.

1. Shiau, L. D., Wen, C. C. and Lin, B. S. Separation and purification of p-xylene from the mixture of m-xylene and p-xylene by distillative freezing, *Ind. Eng. Chem. Res*, **44(7)**, 2258-2265, 2005.

2. Shiau, L. D., Wen, C. C. and Lin, B. S. Separation of p-xylene from the multicomponent xylene system by stripping crystallization, AIChE Journal, 54(1), 337-342, 2008.

350841 Fabrication of Low Cost Insulating Material from Kaolin Clay for Construction Purposes

Abstract:

A new low cost, thermal insulator, construction mineral polymeric material was synthesized from local Saudi Arabia resources using Inorganic Mineral Polymerization technique (MiP). In this technique, the mineral polymer is produced from the following components: clay mineral rich in koalinite, sand, sodium hydroxide solution, and water. Inorganic polymerization experiments are conducted by mixing known amounts of clay minerals and sand in a variable speed stainless-steel mixer for some period of times until a homogenous solid mixture is obtained. Sodium hydroxide solution is added to the mixture and the polymerization reaction takes place at different mixing speeds and times. The produced mixture is divided into a number of samples of known weights. Each sample is moulded in a stainless-steel cylinder at a medium pressure. Then the moulded samples are cured in an oven for 24 hours. The curing temperature was 80°C. The mineral polymerization conditions such as the ratio of clay to sand filler, particle size of the clay, particle size of the sand, concentration and amount of sodium hydroxide, speed and time of mixing were optimized to obtain stable and hard mineral polymer. Compressive strength for different specimens of fabricated mineral polymers was tested under dry and wet conditions. Values of compressive strength with values ranging from 11.5 MPa to 27.9 MPa were obtained.

351141 Thermodynamic Study of Binary PAH (Anthracene + Phenanthrene) Solid Mixtures

Abstract:

Polycyclic aromatic hydrocarbons (PAHs) are common components of many materials, such as petroleum and various types of tars. They are generally present in mixtures, occurring both naturally and as byproducts of fuel processing operations. It is important to understand the thermodynamic properties of such mixtures in order to better understand and predict their behavior (i.e., fate and transport) in the environment and in industrial operations. The Knudsen effusion technique was combined with gas chromatography and differential scanning calorimetry to measure the vapor pressure, composition, and phase behavior of binary anthracene and phenanthrene mixtures at near ambient temperatures. Mixtures of anthracene and phenanthrene exhibited non-ideal mixture behavior: In mixtures with less than 80% phenanthrene, this more volatile component preferentially sublimed and the vapor pressures decreased as the mixtures became enriched in anthracene. The vapor pressure data do not follow Raoult's law (that is, ideal mixture) predictions. Mixtures with high phenanthrene content (> 80%) showed azeotrope like behavior, in that anthracene and phenanthrene sublimed at nearly constant vapor pressure and molecular ratio. These results confirm that the behavior of few-component PAH mixtures does not follow ideal mixture models. Hence, as tars and petroleum (and potentially other organic compound-containing mixtures) weather to few component mixtures over time, complicated, non-ideal phase behavior might be observed.

351160 How Confident Are You That Your Major Accident Risks Are Under Control?

Abstract:

Major accidents in the process industries including those at Texas City , Buncefield UK and in the Gulf of Mexico have highlighted the critical role played by senior managers and executives. Effective leadership is essential if companies are to develop a positive safety culture that remains constantly vigilant towards the risk of a major process safety incident. In the UK the HSE has made leadership a key priority for the high hazard industries and there is a cross-industry initiative to develop and deliver training focussed towards board members responsible for process safety performance.

This paper explores the leadership failings that contributed to recent major accidents, and the cost to both the company reputation and the individuals involved. Key principles are discussed that can help improve leadership in the major hazard industries. These include; attitudes and decisions of senior manager affecting the investment strategy and the safety culture of the organisation, reinforcing the importance of safety by personal example, thorough understanding of major accident hazards and key risk control systems, investigating process safety incident and near misses to find the underlying causes, developing world class safety management systems, and identifying weaknesses in these systems using targeted performance indicators.

The author bases this paper on the development of a training standard for 'Process Safety Leadership' involving a cross-industry 'expert panel'. Training of senior executives in compliance with this standard has been carried out from late 2011 and feedback from these courses and the key requirements for effective leadership will be discussed.

351236 Sil Determination: Shortcomings with the Use of LOPA

Abstract:

Proposed Topic Area: Managing Risks

For a number of years, industry has been trying to apply the principles outlined in IEC 61508. In the process sector, this has resulted in a focus on the requirements of IEC 61511 (2003) and ANSI/ISA 84.00.01 (2004). These are risk based standards. They involve the identification of potential hazardous events, the causes that could lead to each event and the necessary risk reduction for each event. Where the risk reduction involves an instrumented function (e.g. a trip or an interlock), the required performance of the instrumented function can described in terms of the average probability of failure on demand. In the standards, the performance for such instrumented functions is assigned to one of four Safety Integrity Levels. The assessment of what performance level is needed in respect to a specific hazardous event is referred to as SIL Determination. In the field of SIL Determination, IEC 61511 provides examples of a number of different approaches – these are shown in IEC 61511 Part 3. These include Event Trees, Risk Graphs, Safety Matrices, Fault Tree Analysis and Layer of Protection Analysis (LOPA).

In the first few years, following the publication of IEC 61508 Edition 1 in 2000, industry generally embraced the use of Risk Graphs as the method of choice. However, over the subsequent few years, many companies using this method found that it was (a) giving higher safety integrity level targets than really needed and (b) was less flexible than other methods and consequently more difficult to use in many real situations. This has resulted in a general move to other methods and in particular to the use of Layer of Protection Analysis. Indeed, LOPA may now be the predominant "Method of Choice" for SIL Determination. As a method, LOPA gives the impression of simplicity and appears quite straightforward. It almost invites anyone with a minimum of understanding of the method to be a practitioner.

However, there are a number of shortcomings in the use of the LOPA method that have become evident through its use across the globe.

This paper will review the application of LOPA and provide insights into what may be seen as the proper use of the method and its core principles. It will also highlight various shortcomings in use that have been found during expert reviews carried out of a number of existing studies. The paper will be of benefit both to those who are new to the LOPA method, but also of value to those who have been using the method for some time.

353114 Chemical Process Development - a Lean Six Sigma Perspective

Abstract:

Lean Six Sigma (LSS) is a powerful and flexible methodology that has been utilized by a range of companies for delivering significant cost and growth productivity via simultaneous reduction of waste and defects from internal processes. LSS principles have been successfully applied to a full range of both batch and continuous processes in the chemical industry.

This presentation will provide an overview of the application of both the Lean and Six Sigma Methodologies within Momentive, a global manufacturer of specialty chemicals and materials. Several case studies will be shared on how this world class framework has enhanced efforts within our Global Process Development organization.

Using Six Sigma and Experimental Design in Process Development

Mark V. M. Emonds
The Dow Chemical Company

14AIChE Spring Meeting, New Orleans, LA

DOW®

Six Sigma in Process Development

- Six Sigma is frequently used in manufacturing to reduce defects & improve processes
- Why use Six Sigma in process development?
  - Optimize reaction conditions for yield or purity
  - Understand expected variation
  - Produce a more robust process
  - Design a process solution that fits the project needs

14AIChE Spring Meeting, New Orleans, LA

DOW®

Design of Experiments

- Reduce the overall experimental effort
  - More experiments initially, but less overall
- Get to solution quickly, with minimal effort
- Develop an empirical model of the process
- Understand which factors have an impact on process
- Use the model to predict optimal conditions

14AIChE Spring Meeting, New Orleans, LA

DOW®

Type of Model

- Default model is polynomial

  $Y = \beta_0 + \beta_1 x_1 + \beta_2 x_2 + \beta_{12} x_1 x_2 + \beta_{11} x_1^2 + \beta_{22} x_2^2$

- If model form is known based on scientific principles, can incorporate nonlinear relations into experimental design

14AIChE Spring Meeting, New Orleans, LA

DOW®

Keys for Success

- Capable measurement system
  - You cannot study what you cannot measure
- Existing process is repeatable
  - DOE will provide measure of variation, but...
  - If too much variation, any signal lost in the noise
- Acceptable ranges of factors are understood
  - Importance of screening experiments
  - Understand plant/process limitations

14 AIChE Spring Meeting, New Orleans, LA

DOW®

Examples of Application

- Catalyst screening
  - Discovery of new catalysts for new processes
- Process optimization
  - Screen new catalysts for potential improvement
  - Determine optimum reaction conditions
  - Develop empirical model of yield
- Design new solvent system for intermediate
  - Mixture design

14 AIChE Spring Meeting, New Orleans, LA

DOW®

Catalyst Screening

- Use of high throughput screening to search for new catalysts/ligands
  - Initial screening of many catalysts/ligands
  - Use a screening design to detect activity
  - Depending on chemistry, may need to screen under several conditions
    - Kinetics of catalyst A may be too slow at low temp
    - Catalyst B may decompose at higher temperature

14AIChE Spring Meeting, New Orleans, LA

DOW®

Catalyst Screening

- Most promising catalysts or ligands optimized
  - Second round of high throughput screening
  - Factors: activator, concentration, cosolvent
  - Optimize for yield and activity (reaction rate)
- Subsequently validate model in larger scale equipment
  - Verify optimal conditions

14AIChE Spring Meeting, New Orleans, LA

DOW®

High Throughput Screening

- Easy to do large scale experimental designs
- Not always easy or possible to randomize all factors
  - Gas pressure, gas ratios, temperature
- Understand consequences of split-plot designs

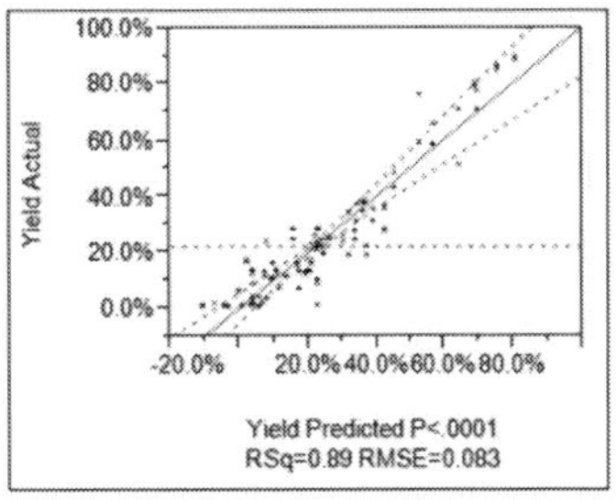

14AIChE Spring Meeting, New Orleans, LA

DOW®

Process Optimization

- Engineering & Process Science group provides support for Amines business
- Understand how new catalysts would perform in transamination reactors
  - Seeking better product selectivity
  - Seeking better productivity
  - Potential for new products

14AIChE Spring Meeting, New Orleans, LA

DOW®

Product Selectivity

- Complex mixture of possible products

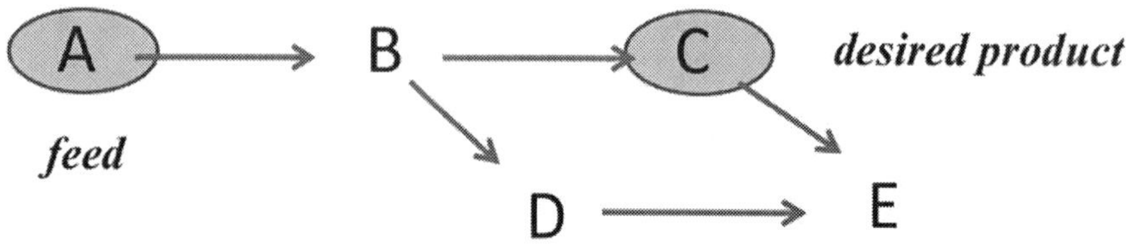

- Want to maximize selectivity and productivity

14AIChE
Spring Meeting, New Orleans, LA

DOW

Comparing New Catalysts

- New catalysts are evaluated in continuous reactor packed bed column
- Are these really different?

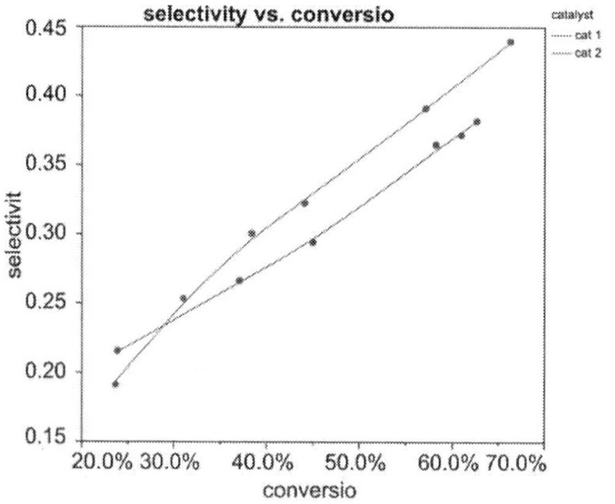

14AIChE
Spring Meeting, New Orleans, LA

DOW

Statistical Analysis

- Fit both runs separately
  - Both fits with $R^2 > 0.99$
- Intercepts of both are same within error
- Slopes are not: outside 95% probability

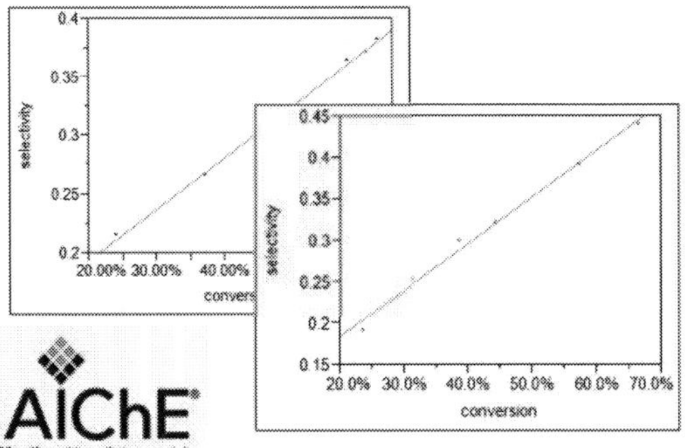

run	cat 1	cat 2
lower 95% interval	0.0848	0.0371
intercept	0.1067	0.0727
upper 95% interval	0.1286	0.1082
lower 95% interval	0.3915	0.4822
slope	0.4353	0.5597
upper 95% interval	0.4791	0.6372

14AIChE Spring Meeting, New Orleans, LA

DOW®

Process Optimization

- After catalyst is selected, further experiments define optimal reaction conditions
  - Feed rates
  - Temperatures
  - Pressures
- Result: a robust model that predicts how the catalyst will perform in the production plant

14AIChE Spring Meeting, New Orleans, LA

DOW®

New Solvent System

- Intermediate for fine chemical synthesis used as a solution to enable stoichiometric control
- Incumbent solvent system not optimal
  - Desired higher concentration of intermediate
    - Current system run at ~ 25% intermediate in solution
  - Used solvent with regulatory issues
  - Solution must be stable and intermediate soluble over a range of temperatures
    - Unheated storage tanks

14AIChE Spring Meeting, New Orleans, LA

Dow®

First DOE

- Optimal design mixture DOE
  - Test solvents existing in process (Solvent A, Solvent B)
  - Determine effect of water levels in solvent
  - Determine if level of stabilizer affects solubility
  - Use Custom Design platform in JMP
  - 19 experiments, including three center points
  - Measured at two temperatures

14AIChE Spring Meeting, New Orleans, LA

Dow®

First DOE Result

- Mixture of Solvent A and Solvent B provide some advantage over single solvent at dissolving intermediate
  - Statistically significant non-linear blending effects
  - Solvent A*Solvent B: Effect Test Prob > F = 0.011
- Water improved maximum solubility
  - Solvent A*Solvent B*water: Prob > F = 0.037
- Stabilizer had no effect

14 AIChE Spring Meeting, New Orleans, LA

DOW®

Model from First DOE

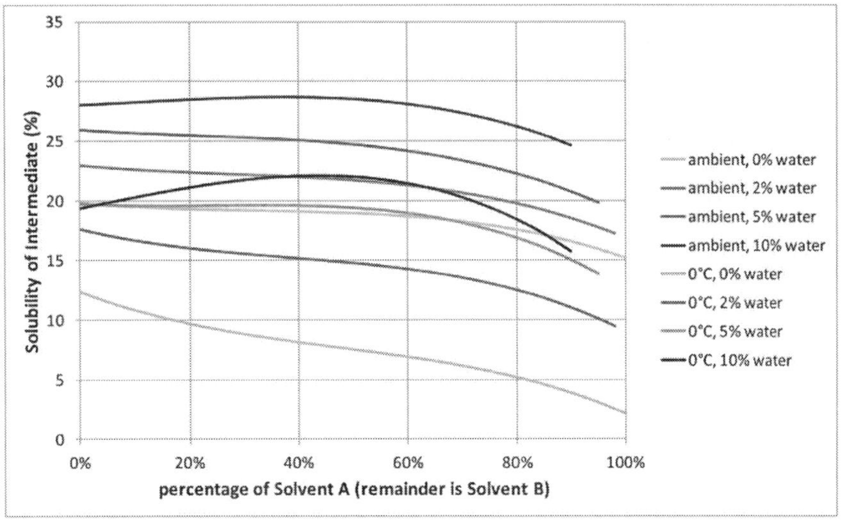

14 AIChE Spring Meeting, New Orleans, LA

DOW®

First DOE Result

- No combination of solvents reached goal
- Maximum modeled solubility at 0°C = 22%

14AIChE Spring Meeting, New Orleans, LA

DOW®

Second DOE

- After screening other solvents, decided to add third solvent to mixture DOE (Solvent C)
  - Solvent C: low cost, compatible with process
- Eliminated stabilizer from experiment
- Optimal design mixture DOE
- 17 experiments, including three center points
- Measured at two temperatures

14AIChE Spring Meeting, New Orleans, LA

DOW®

Second DOE Result

- Significant model
- $R^2 = 0.998$, adj $R^2 = 0.995$
- ANOVA prob > F: < 0.0001

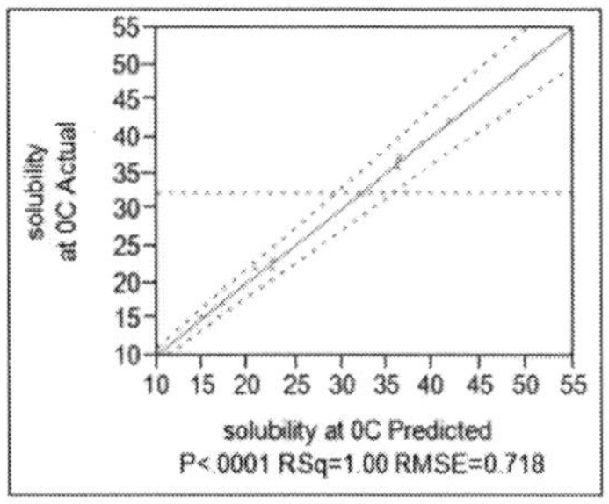

- Solubility modeled separately at each temp
- Statistically significant non-linear blending effects
  - Solvent A*Solvent C: Prob > F of < 0.0001
  - Solvent B*Solvent C: Prob > F of < 0.0001

14AIChE Spring Meeting, New Orleans, LA

Dow®

Model from Second DOE

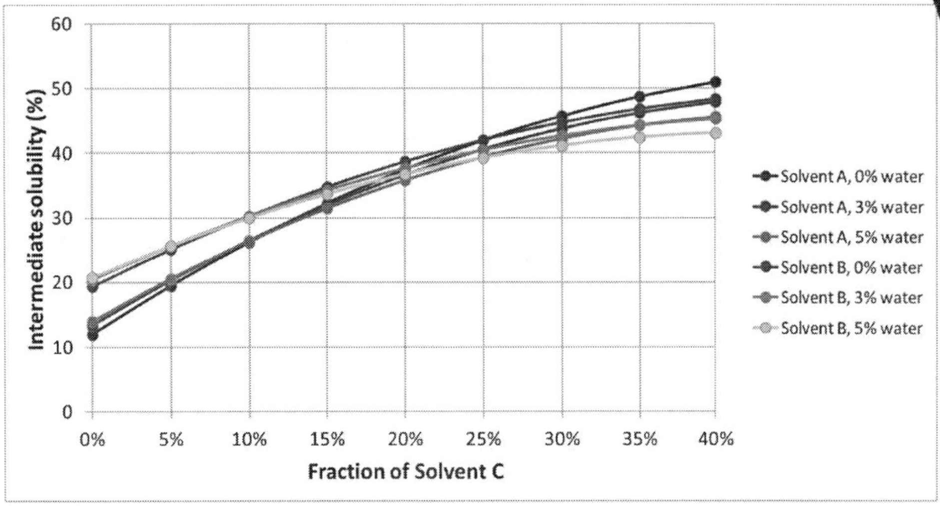

Temperature = 0°C model

14AIChE Spring Meeting, New Orleans, LA

Dow®

Second DOE Result

- Either Solvent A or Solvent B with Solvent C can meet target concentration
- No advantage for three-way mixture
- Did not model with temperature as factor
  - Not required for solution to problem
  - Only collected data at 2 temperatures
  - Temperature statistically significant predictor of solubility, based on analysis of pooled data
  - Other non-linear blending effects

14AIChE Spring Meeting, New Orleans, LA

DOW®

Solvent Optimization Results

- Met target of ≥ 40% intermediate conc.
- Solubility model validated
- Selected solutions tested down to -22°C
- Model provides framework for economic analysis
  - Cost of solvent C
  - Savings from reducing volume of solution
  - Elimination of solvent with regulatory issues

14AIChE Spring Meeting, New Orleans, LA

DOW®

Overall Summary

- Use of Six Sigma can ensure robust, statistically validated solutions
- Optimization of process conditions
- Robust analysis of high throughput data
- Find solutions to complex problems
- Encourage use of Experimental Design in process development

14AIChE
Spring Meeting, New Orleans, LA

DOW®

Lean Six Sigma Methods at Xerox Research Centre of Canada

Dr. Tom Enright
Manager, Process Engineering Area
Xerox Research Centre of Canada

xerox

Xerox Research Centre of Canada (RCC)

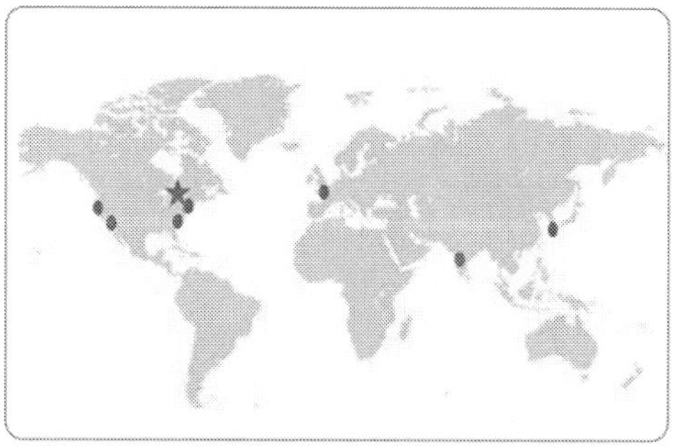

40 years
Materials R&D

110
R&D + Support Staff

>150
Patents Annually

120,000 ft^2
Research Labs & Pilot Plant

Our Vision
Drive business growth by enabling commercialization of breakthrough materials, technologies, and services

2

xerox

Materials R and D at Xerox RCC

Structured Materials Films/Coatings

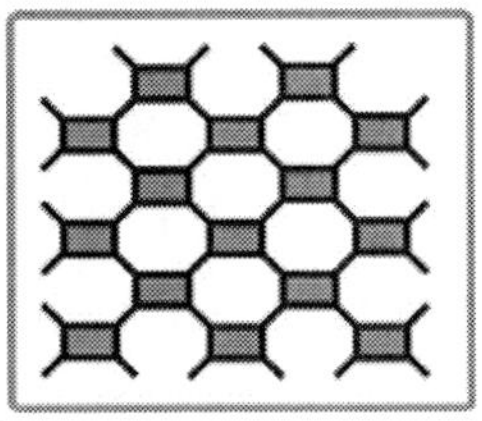

Sustainable Resins

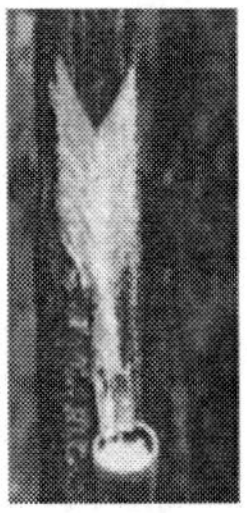

Digital Manufacturing & Electronic Materials

Composites & Nanomaterials

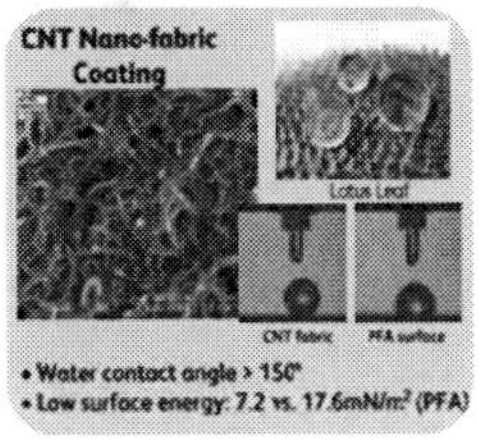

Specialty Materials

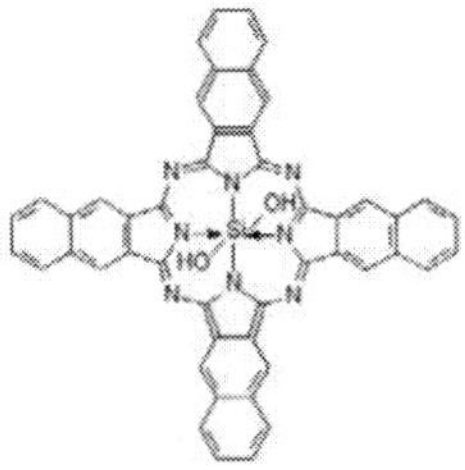

Engineering & Scale-up

From Concept to Commercialization

Xerox RCC Scale-up Engineering and Pilot Plant

I. Process Chemistry

New Materials Design and Synthesis
- Process Chemistry
- QC and Characterization methods
- Performance test development and validation
- Preliminary safety assessment

II. Bench Engineering

Process Scale-Up
- Understand process fundamentals
- Identify Unit Operations
- Optimize process
- Demonstrate process reproducibility and repeatability
- Supply model material lots up to 1 kg
- Define process for piloting
- Deliver preliminary Process Operating Instructions (POI)

III. Pilot Plant

Demonstrate Process Manufacturability
- Equipment selection
- Materials handling/safety
- Process control
- Scaleability demonstration and further optimization
- Large scale sample supply up to tons
- Launch support
- Finalize POI for scale-up to manufacturing
- Technology transfer to manufacturing

IV. Manufacturing

Plant Design & Engineering
- Materials handing at scale
- Process control and automation at scale
- Cost and Quality Control
- Business Continuity assurance

Lean Six Sigma in Xerox Corporation

Lean +	Six Sigma
Speed + Low Cost	Culture + Quality
Lean **Speed** enables Six Sigma Quality (faster cycles of experimentation/ learning)	Six Sigma **Quality** enables Lean Speed (fewer defects means less time spent on rework)

Fast facts about Lean Six Sigma at Xerox

- 1,300 Black Belts trained or certified and active or back in operations
- 9,000 Green Belts trained or certified
- >70% Senior Executives Green Belt trained
- 50,000+ trained in QwikSolver™

For over 10 years we have employed a combination of Lean Six Sigma and Design for Lean Six Sigma methodologies, including:

- **LSS DMAIC** (Define, Measure, Analyze, Improve, Control): A standard approach for re-engineering existing processes
- **DfLSS DMEDI** (Define, Measure, Explore, Develop, Implement): Ideal for designing and optimizing new processes
- **DfLSS IDOV** (Identify, Develop, Optimize, Validate): Used, with specialized tools, for customer driven design of new technologies and services

6 April 8, 2014 xerox

LSS and DfLSS Through the Development Cycle

I. Process Chemistry	II. Bench Engineering	III. Pilot Plant	IV. Manufacturing

New Materials Design and Synthesis	**Process Scale-Up**	**Demonstrate Process Manufacturability**	**Plant Design & Engineering**

DfLSS (IDOV) – New product/process design

DfLSS (DMEDI) – Design and optimize new processes

LSS (DMAIC) – Re-engineer existing processes

xerox

7 April 8, 2014

LSS and DFLSS at Xerox RCC

Design for Lean Six Sigma (DFLSS) / Lean Six Sigma (LSS)

IDOV
Steps to design a new product or service

DMEDI
Steps to design a new product or process

DMAIC
Steps to improve an existing process

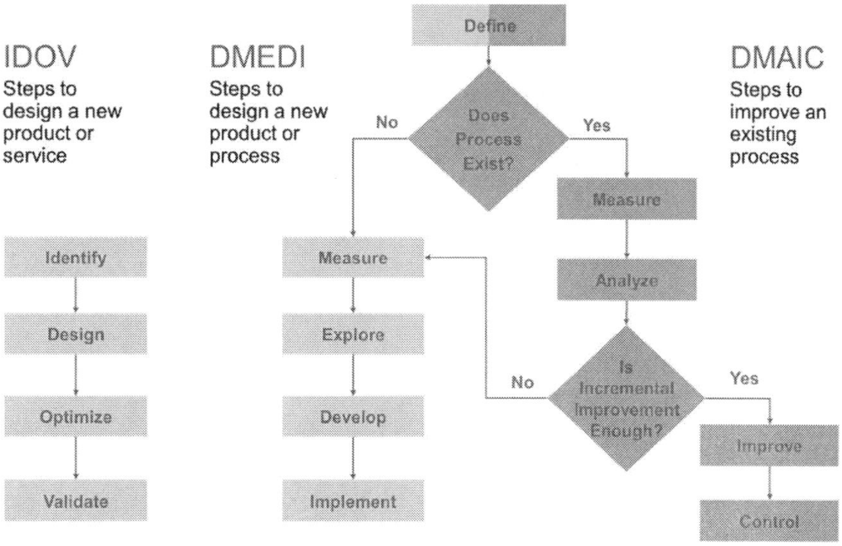

xerox

8 April 8, 2014

Basic Phase Activities

Design for Lean Six Sigma (DFLSS)

Identify		Design	Optimize	Validate
Define	Measure	Explore	Develop	Implement
• Define the opportunity from both business and customer perspectives	• Translate Voice of the Customer into Critical Customer Requirements	• Develop design concepts and high-level design	• Develop and optimize detailed design	• Validate design with pilot, develop control plan, implement

Lean Six Sigma (LSS)

Define	Measure	Analyze	Improve	Control
• Establish problem or opportunity statement • Establish success criteria	• Establish current state, identify where a process can be streamlined	• Identify root causes • Identify time intensive activities of a process • Identify how to eliminate nonvalue-added steps	• Develop, plan solutions • Run pilot • Failure analysis	• Implement solution • Monitor success of solutions • Develop process controls and mistake proofing

LSS and DFLSS Roles at Xerox RCC

- Nine Black Belt certified/trained staff
  - Mixture of DMAIC, DMEDI, DfLSS
  - Forty hours online plus five weeks in-class training
  - Lead black belt projects and demonstrate tool usage
  - Specialized skill training (e.g., Shainin, Kaizen, TRIZ)
  - Ability to select appropriate project path, lead projects, coach project leaders
  - At least one Black Belt assigned to every Xerox RCC project
- Remaining technical staff are Green Belt certified (DfLSS)
  - 40 hours online plus one week in-class training
  - Studies and tool demonstration
  - Good understanding of common tools with some specialization
- Access to Xerox corporate Master Black Belt coaches for complex projects

DfLSS (IDOV) Roadmap

Identify
- Project Charter
- Strategic Plan
- Cross-Functional Team
- Voice of the Customer
- Benchmarking
- KANO's Model
- Questionnaires
- Focus Groups
- Interviews
- Internet Search
- Historical Data Analysis
- Design of Experiments
- Quality Function Deployment
- Pairwise Comparison
- Analytical Hierarchy Process
- Performance Scorecard
- Flow Charts
- FMEA
- Visualization

Develop
- Assign Specifications to CTC's
- Customer Interviews
- Formulate Design Concepts
- Pugh Concept Generation
- TRIZ or ASIT
- FMEA
- Fault Tree Analysis
- Brainstorming
- QFD
- Scorecard
- Transfer Function
- Design of Experiments
- Deterministic Simulators
- Discrete Event Simulation
- Confidence Intervals
- Hypothesis Testing
- MSA
- Computer Aided Design
- Computer Aided Engineering

Optimize
- Histogram
- Distributional Analysis
- Empirical Data Distribution
- Expected Value Analysis (EVA)
- Adding Noise to EVA
- Non-Normal Output Distributions
- Design of Experiments
- Multiple Response Optimization
- Robust Design Development
  - Using S-hat Model
  - Using Interaction Plots
  - Using Contour Plots
- Parameter Design
- Tolerance Allocation
- Design For Manufacturability and Assembly
- Mistake Proofing
- Product Capability Prediction
- Part, Process, and SW Scorecard
- Risk Assessment
- Reliability
- Multidisciplinary Design Optimization (MDO)

Validate
- Sensitivity Analysis
- Gap Analysis
- FMEA
- Fault Tree Analysis
- Control Plan
- PF/CE/CNX/SOP
- Run/Control Charts
- Mistake Proofing
- MSA
- Reaction Plan
- High Throughput Testing

11 April 8, 2014

xerox

DfLSS DMEDI Roadmap

Define
- Define the Opportunity
- Complete Charter
- Develop Project Plan
- Begin Risk Management Strategy
- Develop Communication Plan
- Finalize Project Focus

Measure
- Understand the Voice of the Customer
- Translate the VOC to CCRs
- Develop SIPOC Chart
- Develop Process Map
- Analyze Existing Data
- Implement Quick Wins
- Validate Business Opportunity
- Create a Scorecard

Explore
- Generate high-level concept ideas
- Select the concept that best meets VOC & VOB
- Begin high-level design elements
- Create QFD House #2
- Analyze existing data
- Proactively manage risk

Develop
- Develop detailed design
- Create detailed design elements
- Design for Lean
- Optimize design performance
- Confirm design capability

Implement
- Plan and execute a Pilot
- Develop Process Control Plan
- Finalize Documentation
- Implement Solution
- Transition Project to Process Owner
- Identify Project Replication Opportunities
- Calculate Financial Benefits

Define
- Project Selection Tools
- PIP Management Process
- Financial Analysis
- Project Charter
- Multi-Generation Plan
- Project Planning
- Gantt Chart
- RACI
- Risk Matrix
- Communication Plan

Measure
- Kano Model
- Interviews, Focus Groups, & Surveys
- VOC Plan
- Operational Definitions
- Affinity Diagram
- Structure Tree
- QFD House #1
- SIPOC, Top Down & Swim-Lane Process Maps
- Non-Value Added Analysis
- Basic Statistics/Histogram
- Pareto Chart
- Box Plot
- Kaizen Event
- Scorecard
- Target Costing

Explore
- Application Analysis
- Functional Analysis
- Creativity Tools
- Brainstorming
- Benchmarking
- Pugh Matrix
- Pairwise Comparisons
- Analytic Hierarchy Process
- QFD House #2
- Capability Analysis
- Statistical Tolerancing
- Queuing Theory
- Crystal Ball
- Hypothesis Testing
- FMEA
- ANOVA
- Scorecard

Develop
- Value-Added Analysis
- Mistake Proofing/Poka-Yoke
- 5S
- Line Balancing
- Work Control Systems
- Design of Experiments (DOE)
- Full and fractional factoria designs
- Main effects plot
- Interaction plot
- Conjoint Analysis
- FMEA
- To-Be Process Maps
- Risk Assessment Tools
- Scorecard

Implement
- Process Control Plan
- QFD House #3
- Control Charts
- Standard Operating Procedures
- Training Plan
- Communication Plan
- Implementation Plan
- Visual Process Control
- Process Control Plans
- Project Commissioning
- Scorecard
- Project Replication
- Financial Analysis Tools

12 April 8, 2014

xerox

LSS DMAIC Roadmap

	Define	Measure	Analyze	Improve	Control
Activities	• Identify Problem • Complete Charter • Develop SIPOC Map • Map Business Process • Map Value Stream • Gather Voice of the Customer & Voice of the Business • Develop CCR's & CBR's • Finalize Project Focus	• Identify Key Input, Process and Output Metrics • Develop Operational Definitions • Develop Data Collection Plan • Validate Measurement System • Collect Baseline Data • Determine Process Performance/Capability • Validate Business Opportunity	• Propose Critical X's • Prioritize Critical X's • Conduct Root Cause Analysis on Critical X's • Validate Critical X's • Estimate the Impact of Each X on Y • Quantify the Opportunity • Prioritize Root Causes	• Develop Potential Solutions • Develop Evaluation Criteria & Select Best Solutions • Evaluate Solution for Risk • Optimize Solution • Develop 'To-Be' Process Map(s) and High-Level Implementation Plan • Develop Pilot Plan & Pilot Solution	• Develop SOP's, Training Plan & Process Control System • Implement Process Changes and Controls • Monitor & Stabilize Process • Transition Project to Process Owner • Identify Project Replication Opportunities • Calculate Financial Benefits
Tools	• Pareto Charts • Project Selection Tools • PIP Management Process • Value Stream Map • Various Financial Analysis • Charter Form • Stakeholder Analysis • Communication Plan • SIPOC Map • High-Level Process Map • Non-Value Added Analysis • VOC and Kano Analysis • RACI and Quad Charts	• SIPOC Map • Operational Definitions • Data Collection Plan • Statistical Sampling • Measurement System Analysis (MSA), Gage R&R • Constraint Identification • Setup Reduction • Generic Pull • Kaizen • TPM • Control Charts • Process Capability, Cp & Cpk	• Pareto Charts • C&E Matrix • C&E/Fishbone Diagrams • Brainstorming • Detailed 'As-Is' Process Maps • Basic Statistical Tools • SupplyChainAccelerator Analysis • Non Value-Added Analysis • Hypothesis Testing • FMEA • Box Plots • Interaction Plots • Simple & Multiple Regression • ANOVA	• Brainstorming • Benchmarking • Process Improvement Techniques • Line Balancing • Process Flow Improvement • Replenishment Pull • Purchasing and Sales Strategy • Poka-Yoke • FMEA • Solution Selection Matrix • 'To-Be' Process Maps • Piloting and Simulation	• Control Charts • Standard Operating Procedures (SOP's) • Training Plan • Communication Plan • Implementation Plan • Visual Process Control • Process Control Plans • Project Commissioning • Project Replication • Plan-Do-Check-Act Cycle

DMAIC Case Study – Silver Nanoparticle Inks

- Internal Xerox RCC research project for several years to develop silver nanoparticle inks for electronic applications
  - Designed around general material properties, not directed towards specific customer application
- Now selling inks to external clients for specialty applications
  - Customer requirements started to become more clear
  - Inks require adjustments to meet requirements for different applications
- Project path: DMAIC project to adjust material properties into desired ranges

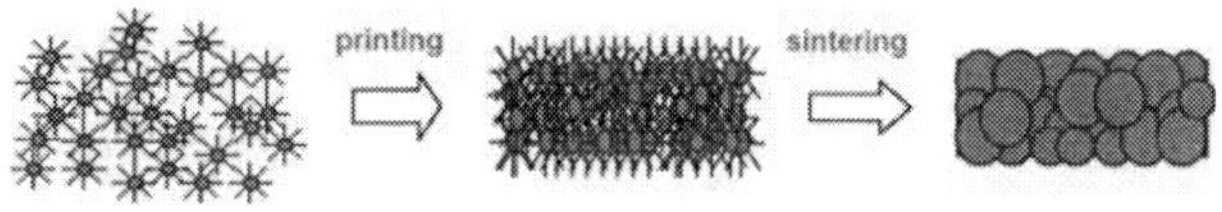

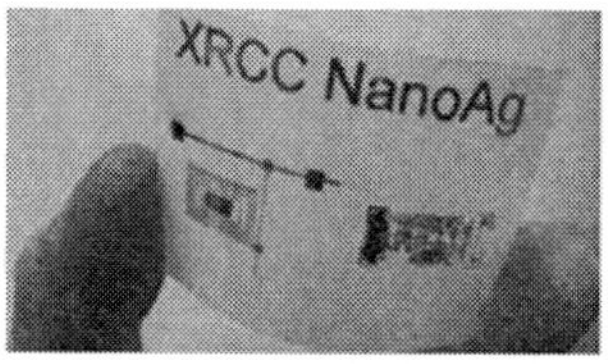

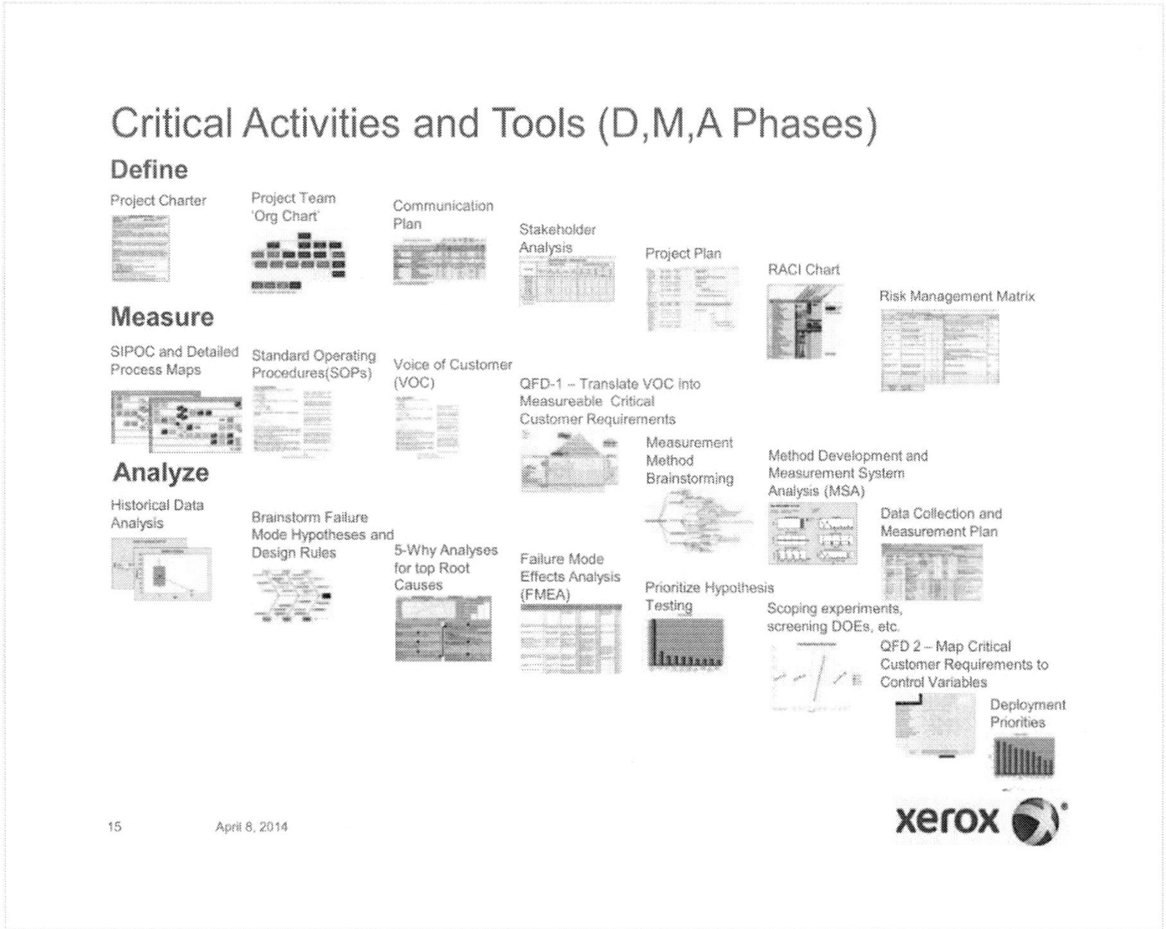
Critical Activities and Tools (D,M,A Phases)
Define
Project Charter
Project Team 'Org Chart'
Communication Plan
Stakeholder Analysis
Project Plan
RACI Chart
Risk Management Matrix
Measure
SIPOC and Detailed Process Maps
Standard Operating Procedures(SOPs)
Voice of Customer (VOC)
QFD-1 – Translate VOC into Measureable Critical Customer Requirements
Measurement Method Brainstorming
Method Development and Measurement System Analysis (MSA)
Analyze
Historical Data Analysis
Brainstorm Failure Mode Hypotheses and Design Rules
5-Why Analyses for top Root Causes
Failure Mode Effects Analysis (FMEA)
Prioritize Hypothesis Testing
Data Collection and Measurement Plan
Scoping experiments, screening DOEs, etc.
QFD 2 – Map Critical Customer Requirements to Control Variables
Deployment Priorities
15
April 8, 2014
xerox

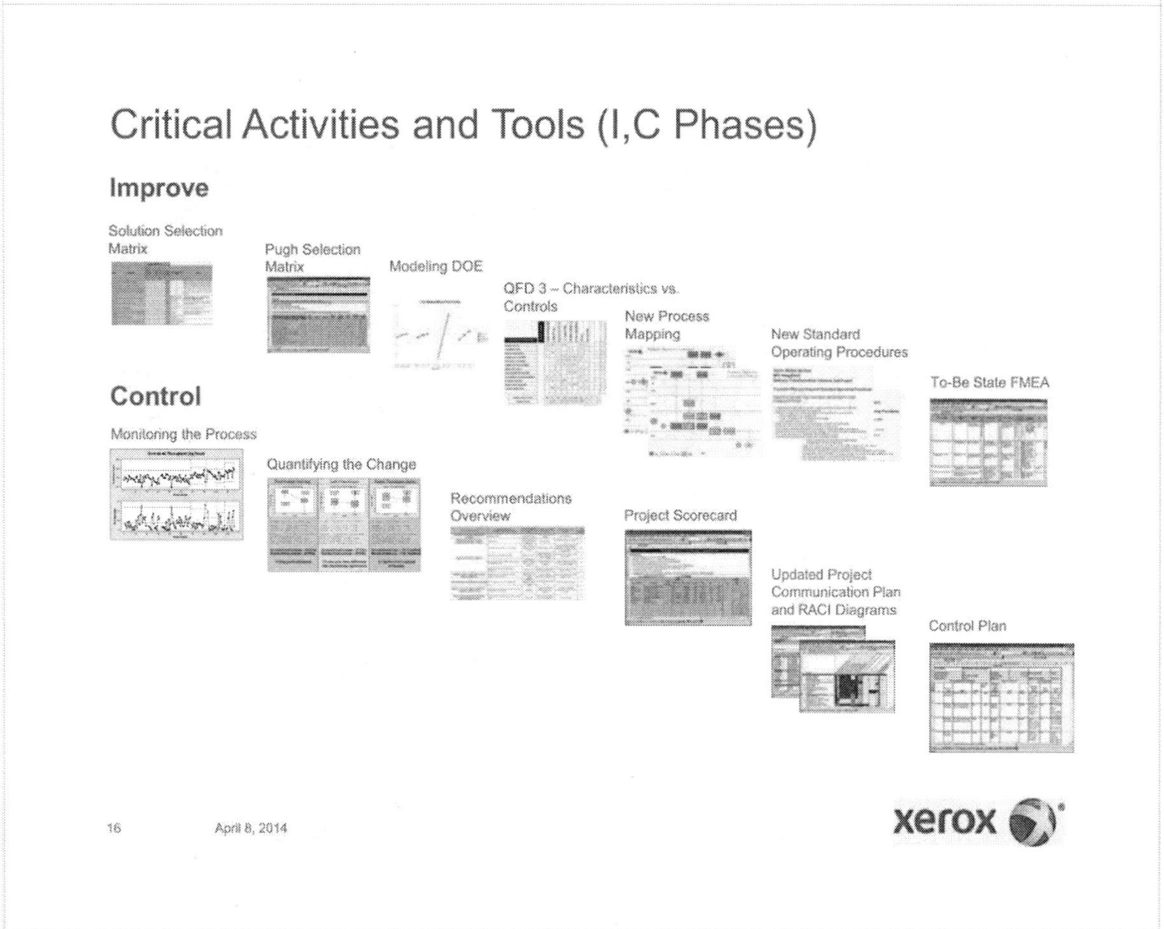
Critical Activities and Tools (I,C Phases)
Improve
Solution Selection Matrix
Pugh Selection Matrix
Modeling DOE
QFD 3 – Characteristics vs. Controls
New Process Mapping
New Standard Operating Procedures
To-Be State FMEA
Control
Monitoring the Process
Quantifying the Change
Recommendations Overview
Project Scorecard
Updated Project Communication Plan and RACI Diagrams
Control Plan
16
April 8, 2014
xerox

Summary

- Xerox RCC uses a flexible LSS/DFLSS methodology to manage projects through the development cycle:
  - IDOV for designing new technologies and processes
  - DMEDI for designing and optimizing new processes
  - DMAIC for improvement of existing processes or products
  - Series of other tools available for smaller activities
    - QuikSolver, Kaizen, etc.
- Rigorous front-end project focus with solid documentation and follow through ensures smoother project delivery cycle
- Proving to be especially effective for managing Client Services engagements

QuikSolver™

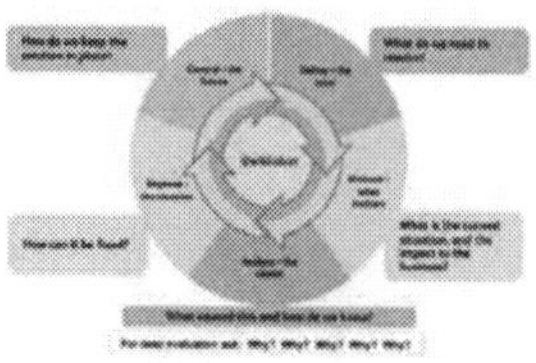

Project Delivery Cycle

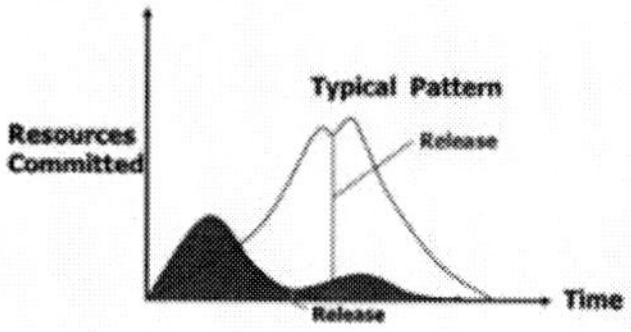

353102 Development of Continuous Improvement Network in Xerox Manufacturing

Abstract:

Xerox Corporation has five North American manufacturing sites that produce toner/developer materials for a wide variety of devices. For many years, each site worked independently and there was no formal mechanism for sharing best practices, successes, and failures. There were attempts over the years of developing some type of information sharing network, but a sustainable process never really took hold for a number of reasons.It was recently decided to make another attempt at developing a formal Continuous Improvement Network through a Lean Six Sigma Black Belt project with team members from each of the manufacturing units. This talk will give an overview of steps that were taken in the project to formulate a robust information sharing network that has checks and balances included to ensure that the network will last in the long term.

352183 Use of Six Sigma Tools in the Commercial Scale-up of a Novel PPE Copolymer

Abstract:

Poly(phenylene ether) is an engineering thermoplastic known for its excellent hydrolytic stability, dimensional stability and inherent flame retardancy. A block copolymer produced by the oxidative coupling polymerization of 2,6 xylenol and a comonomer, has been developed to deliver significant improvements in mechanical properties and flame retardancy.

Process development for new technology to culminate in production in an existing operating commercial chemical plant requires conscientious planning and risk mitigation efforts. This presentation illustrates the evolution of this commercial scale up effort as a Six Sigma project. Special emphasis is done on the application of Six Sigma tools and their impact on the planning of the project requirements, risk mitigation, and implementation analysis and control.

352632 A Qualitative in Field Test to Characterize Oil-Shale Reservoirs

Abstract:

Shale oil recovery, despite its extraction costs and low yield per shale matrix is becoming a more feasible solution to the rapidly emerging global fuel demands. However, many oil-shale reserves are located several thousand feet below the earth's surface, which makes it extremely difficult and expensive to extract. In addition, the quality of the shale matrix may have a broad range such that the potential yield through and other methods may vary significantly across neighboring reservoirs. And determining the quality of the oil-shale matrix and its potential yield of synthetic/tight oil using laboratory analytical methods can take several weeks. This work presents a rapid and simple qualitative "in-field" test-tube test method that categorizes and prioritizes a given shale matrix based upon its composite. This test method can be a significant cost benefit and a proof positive indicator of where to focus exploration and recovery interest.

Pressure Regulation and Kick Attenuation with Wired Drill Pipe Technology in Managed Pressure Drilling

Hector Perez
Reza Asgharzadeh Shishavan
John Hedengren
Brigham Young University

Outline

- Overview
- Motivation for Pressure Control
- Method and Structure of the Control System
- Results
- Results of Automated BHA Pressure

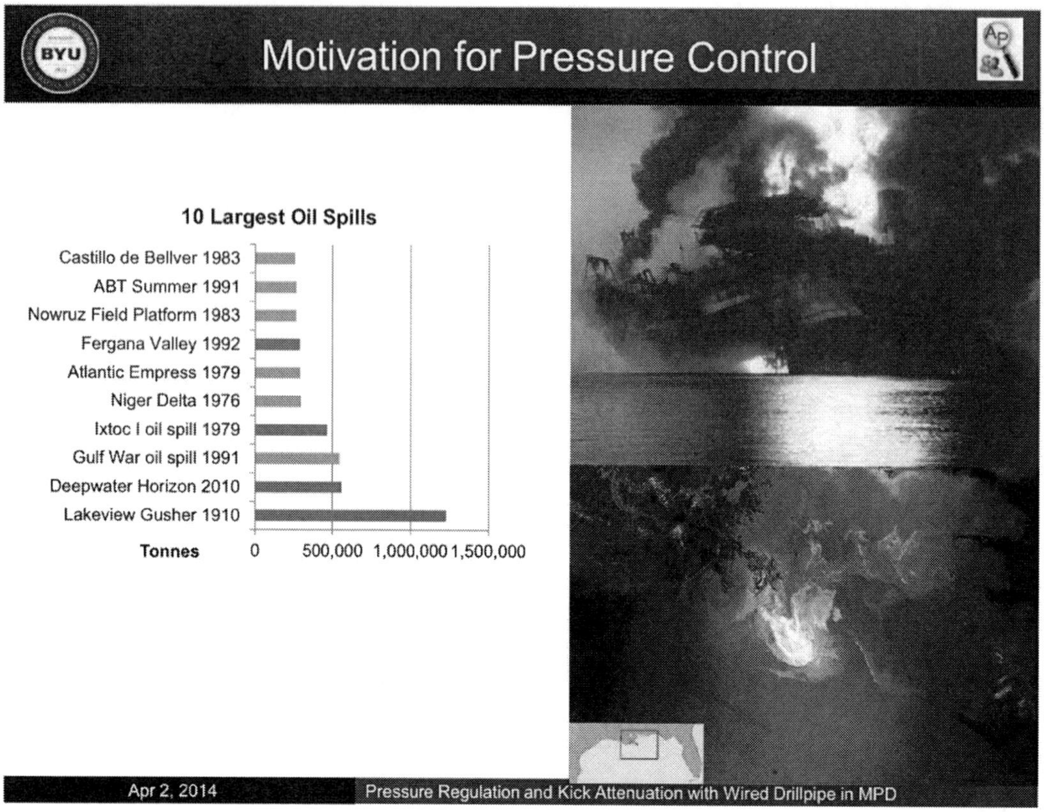
Motivation for Pressure Control
10 Largest Oil Spills
Castillo de Bellver 1983
ABT Summer 1991
Nowruz Field Platform 1983
Fergana Valley 1992
Atlantic Empress 1979
Niger Delta 1976
Ixtoc I oil spill 1979
Gulf War oil spill 1991
Deepwater Horizon 2010
Lakeview Gusher 1910
Tonnes
0
500,000
1,000,000
1,500,000
Apr 2, 2014
Pressure Regulation and Kick Attenuation with Wired Drillpipe in MPD

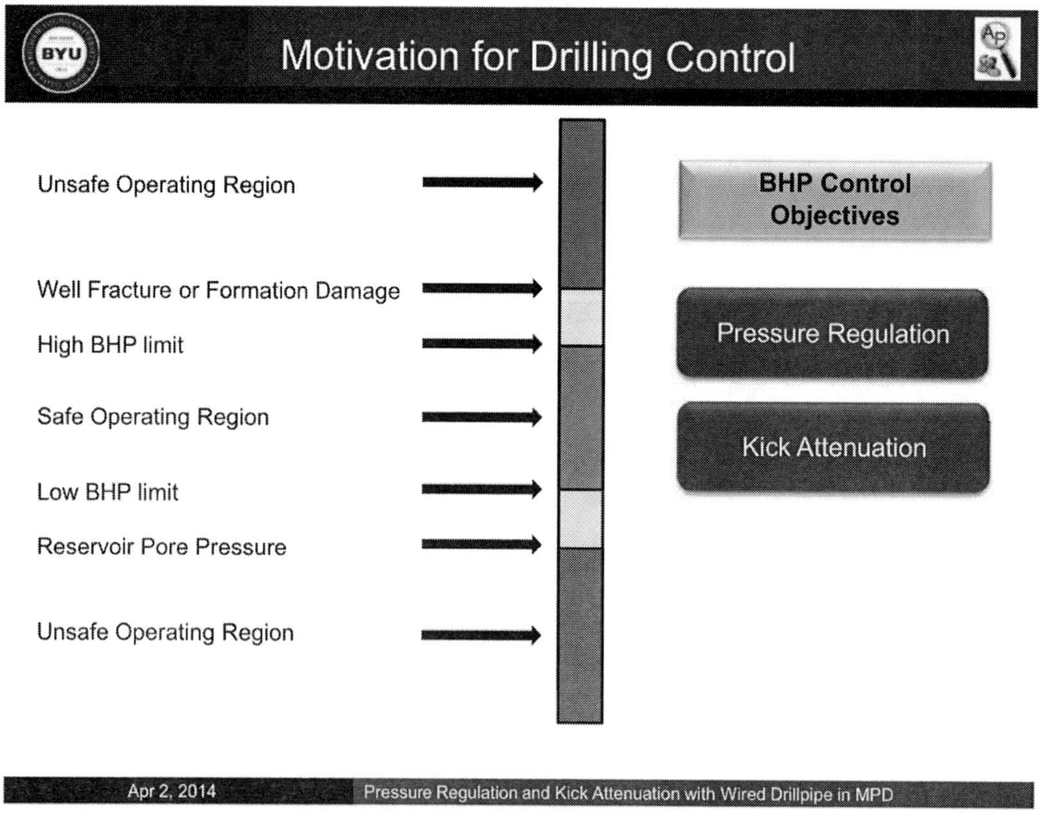
Motivation for Drilling Control
Unsafe Operating Region
Well Fracture or Formation Damage
High BHP limit
Safe Operating Region
Low BHP limit
Reservoir Pore Pressure
Unsafe Operating Region
BHP Control Objectives
Pressure Regulation
Kick Attenuation
Apr 2, 2014
Pressure Regulation and Kick Attenuation with Wired Drillpipe in MPD

Managed Pressure Drilling System

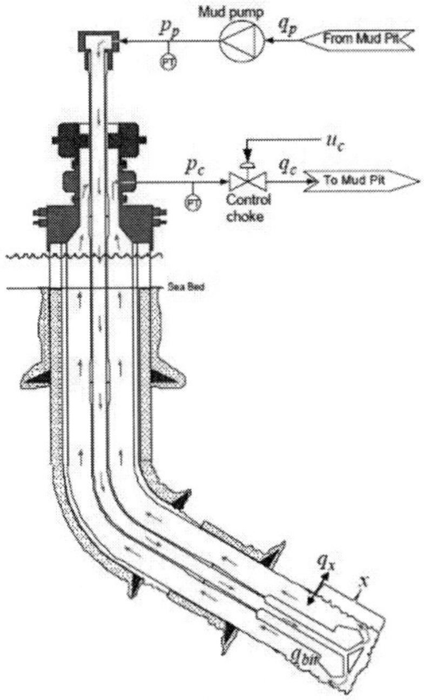

- ❑ Overbalanced Drilling
- ❑ Managed Pressure Drilling (MPD)
- ❑ Underbalanced Drilling

MPD offers:

- ❑ More precisely control of borehole pressure
- ❑ Decrease NPD and mitigate drilling hazards

Manual Control versus Automated Control

Challenges

- Large Time Delays
- Complex Behavior of Multiphase flow
- Sudden Change in Formation Properties

Benefits of BHP control Automation

- Improve Reliability of Pressure control
- Reduce the possibility of human errors

Objective

Designing a comprehensive controller which Simultaneously

- Regulates the BHA Pressure within specified range
- Attenuates Unwanted gas influx when it happens
- Uses latest automation technologies
- Has improved performance in comparison with manual and conventional controllers

Wired Pipe Technology to be used in Control Scheme

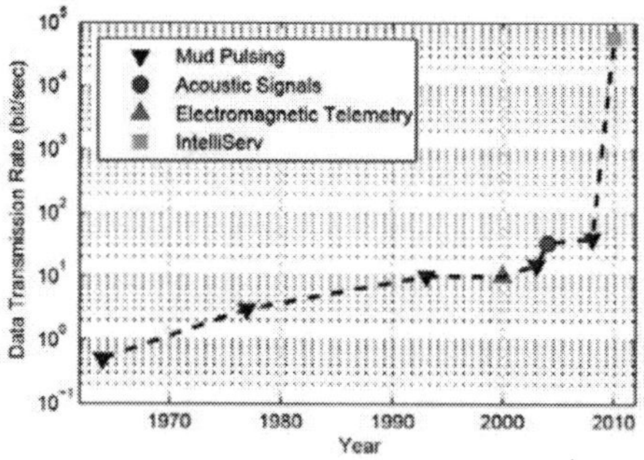

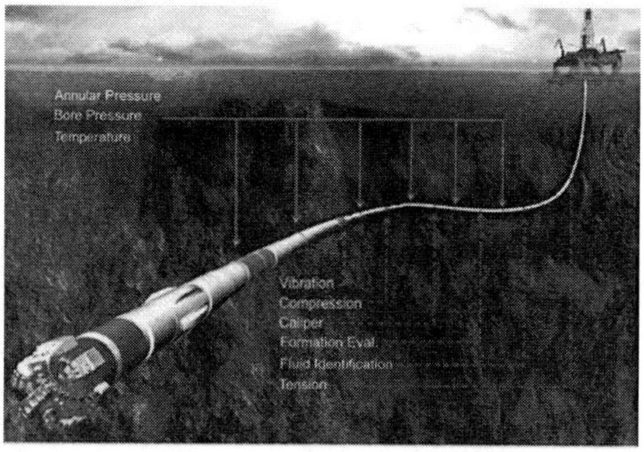

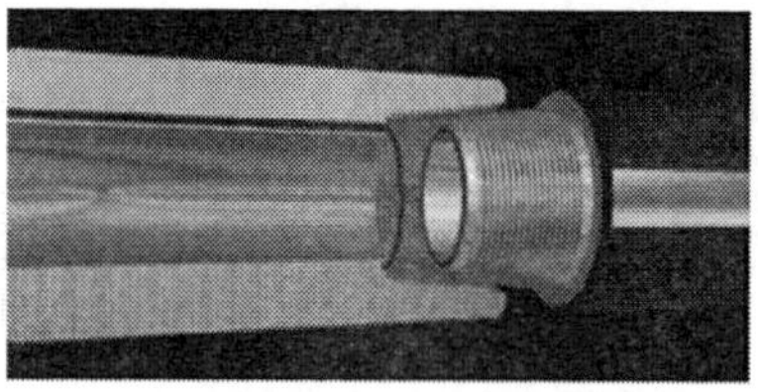

Challenges in the designed controllers

Mud Pulse Telemetry Based

- ❑ Indirect control of BHP
- ❑ High uncertainty in estimated variables values
- ❑ Still Large time delay in kick detection

Wired Pipe Telemetry Based

- ❑ Directly control of BHP and annular pressures ✔
- ❑ Increased precision in estimating the unmeasured process variables ✔
- ❑ Improved kick detection ✔

Schematic of the MPD Controller

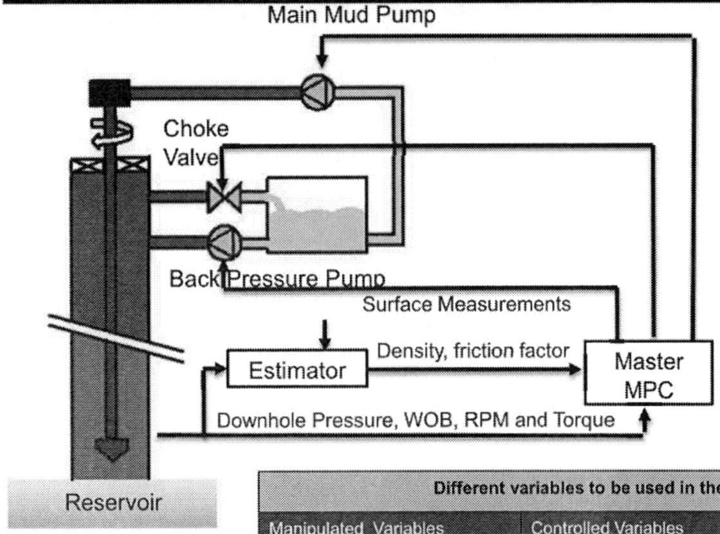

Different variables to be used in the current research		
Manipulated Variables	Controlled Variables	Estimated Variables
Rig Pump Flow Rate	BHP	Annulus Drill Fluid Density
Choke Valve Opening	Annular Pressures	Annulus Friction Factor
		Reservoir Pore Pressure

Controller Structure

Multivariate Nonlinear Model Predictive Controller (MPC)

What are advantages over conventional controllers?

Nonlinear Feature of the Controller

Multivariate Nonlinear Model Predictive Controller (MPC)

- Drilling is a non linear process
- Current controllers use linear (simplified) models
- A nonlinear controller can capture the dynamics better
- Designing nonlinear controllers are more challenging

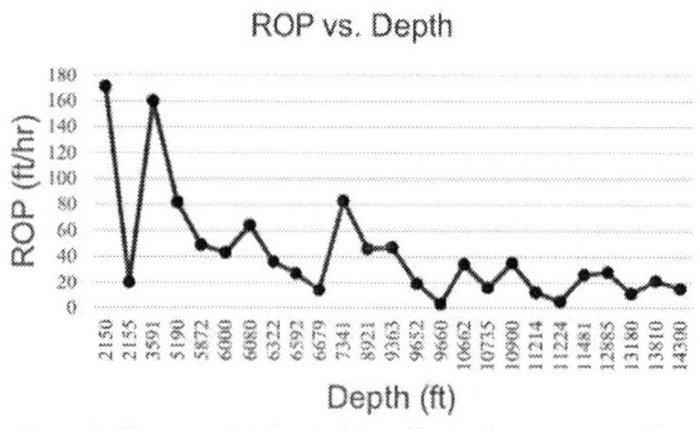

Figure 1. Change of ROP with drilling depth. Field data from (Kinabalu East-1, 1990)

Model Predictive Control and Moving Horizon Estimator

Controller

Multivariate Nonlinear Model Predictive Controller (MPC)

- Future prediction capability – Improved control performance
- MPC vs. Conventional Controllers

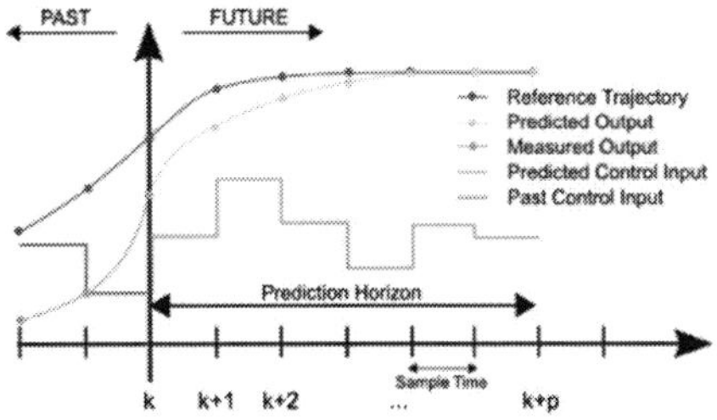

Estimator

- Moving Horizon Estimator
- Uses both downhole and surface measurements
- More precise and faster parameter and state estimation

Software

- APMonitor Optimization Suite
- Designed Especially for Dynamic Nonlinear Processes
- Using the Most Efficient Solvers Available
- Solution Time: Fraction of a Second

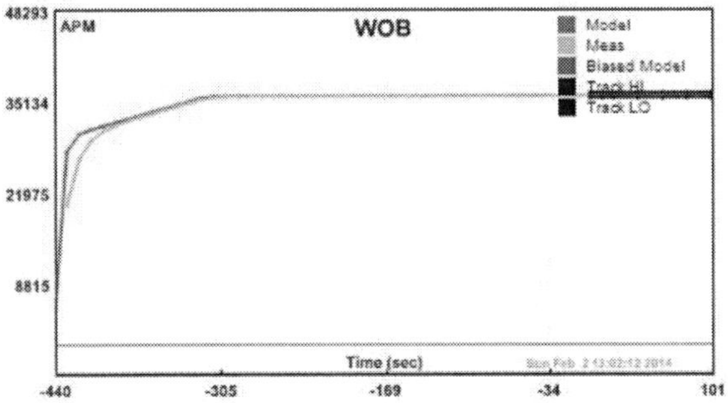

Lower Order Model

Balance Equations

$$\dot{p}_p = \frac{\beta_d}{V_d}(q_{pump} - q_{bit})$$

$$\dot{p}_c = \frac{\beta_a}{V_a}(q_{bit} + q_{back} - q_{choke} + q_{res})$$

$$\dot{q}_{bit} = \frac{1}{M}(p_p - F_d|q_{bit}|q_{bit} + \rho_d g h_{bit} - p_{bit})$$

$$\dot{\mathbf{h}}_{bit} = \boldsymbol{ROP}$$

Closure Relations

$$M_a = \rho_a \int_0^{l_w} \frac{1}{A_a(x)} dx$$

$$M_d = \rho_d \int_0^{L_{dN}} \frac{1}{A_d(x)} dx$$

$$M = M_a + M_d$$

$$p_{bit} = p_c + \rho_a \times F_a|q_{bit} + q_{res}|(q_{bit} + q_{res})h + \rho_a g(h_{bit} - (n-1)h)$$

$$q_{choke} = K_c z_{choke}\sqrt{\rho_{an}(p_c - p_0)}$$

Zhou, Kaasa, et al. (2008)

Apr 2, 2014 Pressure Regulation and Kick Attenuation with Wired Drillpipe in MPD

Kick Detection and Attenuation

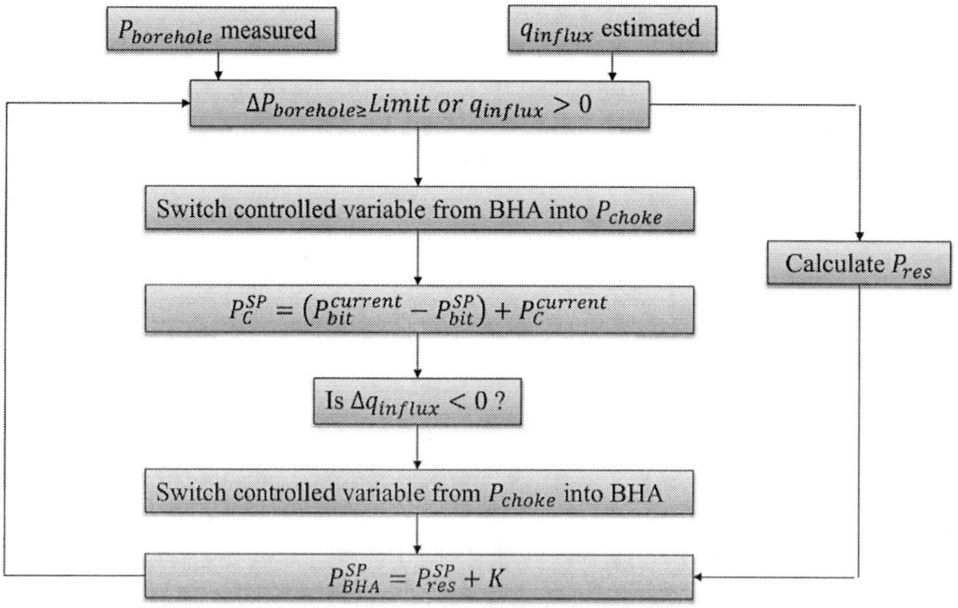

Apr 2, 2014 Pressure Regulation and Kick Attenuation with Wired Drillpipe in MPD

Results

Apr 2, 2014 Pressure Regulation and Kick Attenuation with Wired Drillpipe in MPD

Automated BHA pressure control

Table 1. Initial well conditions used for simulations

Variable	Value (SI units)	Value (US units)
Distance between two successive pressure sensors h_a	150 m	492.1 ft
ROP	0.01 m/s	0.033 ft/s
Fluid density in the drill string - ρ_d	0.012 x 10^{-5} kg/m³	74.9 lb/ft³
Fluid density in the annulus - ρ_a	0.014[a] x 10^{-5} kg/m³	87.4 lb/ft³
Bulk modulus in the annulus β_d	14000 bar	2.03 x 10^5 psi
Bulk modulus in the annulus β_a	14000 bar	2.03 x 10^5 psi
Annulus cross sectional area	0.021 m²	0.226 ft²
Friction factor in the drill string F_d	10000 bar m^{-6} s^2	116.3 psi ft^{-6} s^2
Friction factor in the annulus F_a	1250[a] m^{-5}	3.29 ft^{-5}
Lumped density per length parameter for the drill string - M_d	5700 x 10^{-5} kg/m⁴	108.5 x 10^{-5} lb/ft⁴
Lumped density per length parameter for the annulus – M_a	800 x 10^{-5} kg/m⁴	15.22 x 10^{-5} lb/ft⁴
Well depth (h_{bit})	2000 m	6562 ft
[a]For connection procedures, the values used for simulation were 0.01 (density) and 1000 (friction factor)		

Apr 2, 2014 Pressure Regulation and Kick Attenuation with Wired Drillpipe in MPD

MHE Estimator

Within 30 seconds of operation, the estimator has accurately determined the friction factor and has determined the annulus density within 1.5% of the actual value.

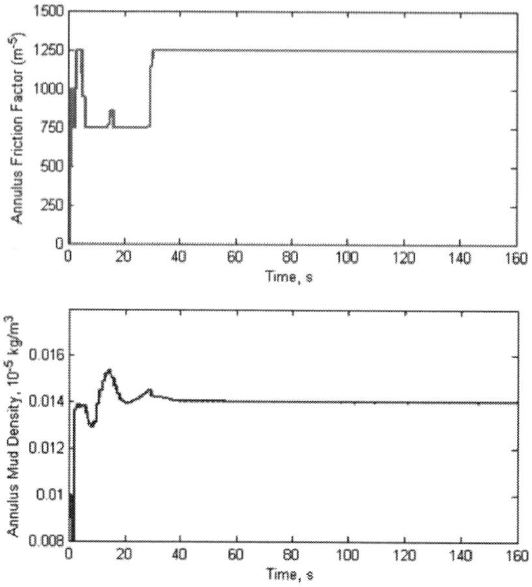

Case 1: Normal Operation

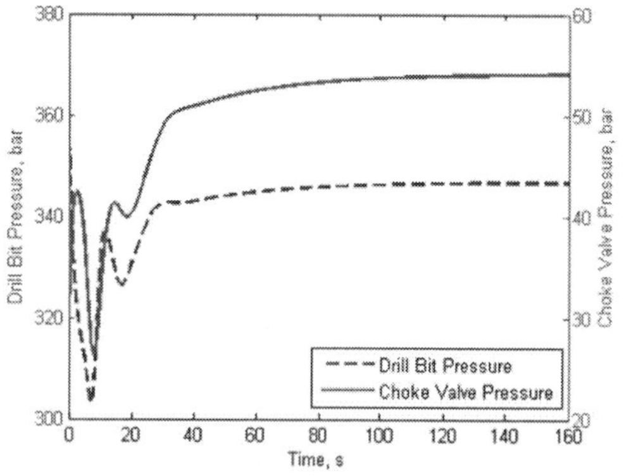

Within 30 seconds, the drill bit pressure has been controlled within 0.72% of its set point of 345 bar

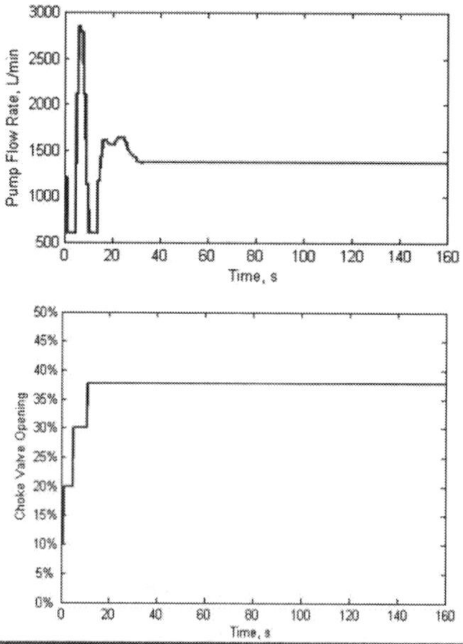

Case 2: Connection Procedure

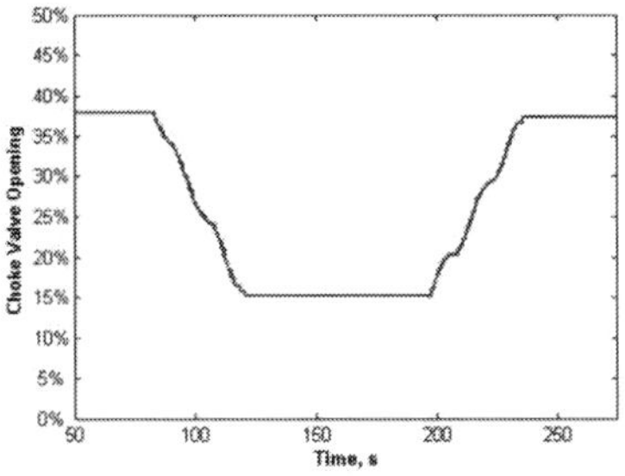

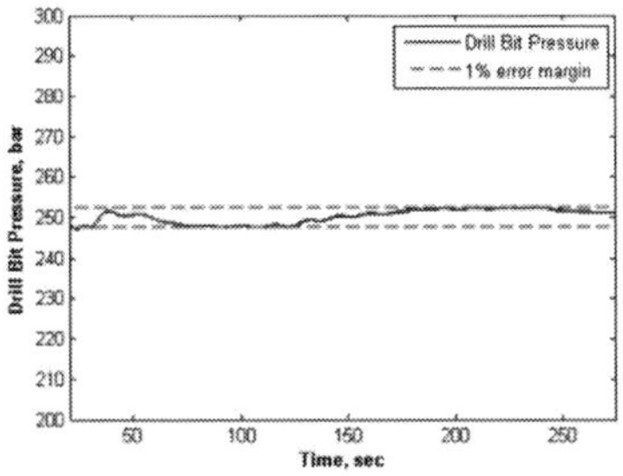

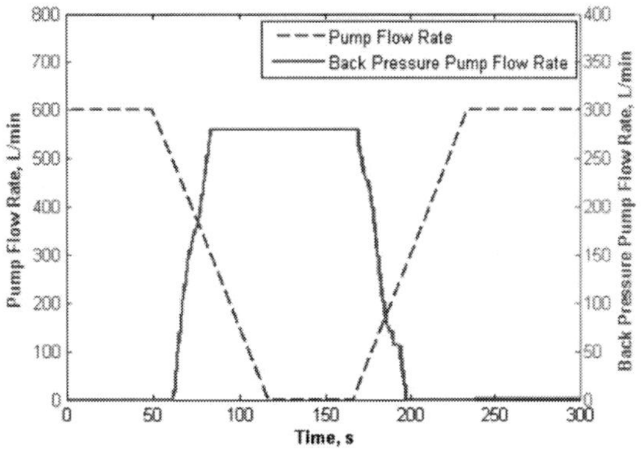

Apr 2, 2014 | Pressure Regulation and Kick Attenuation with Wired Drillpipe in MPD

Case 3: Unwanted Gas Influx

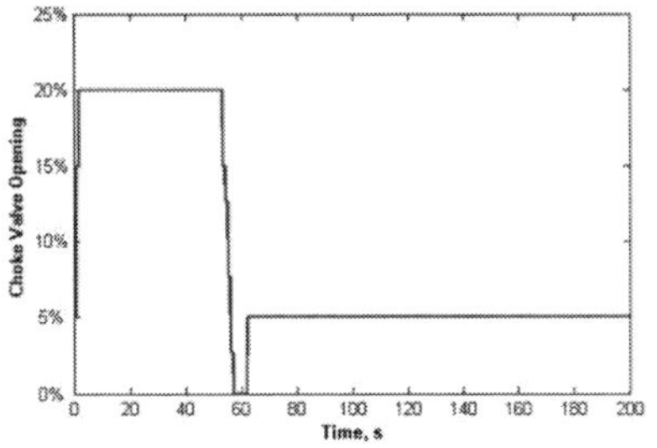

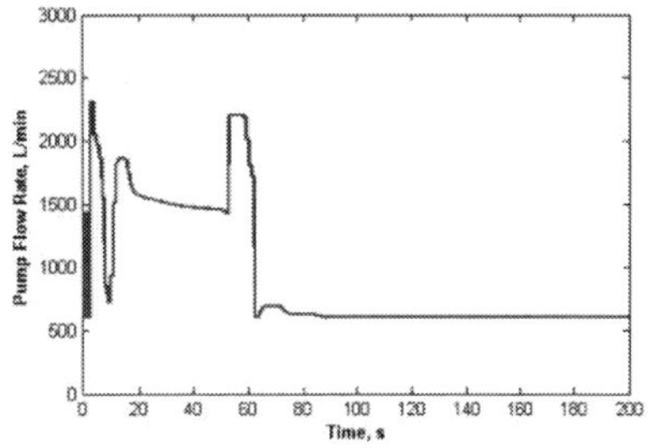

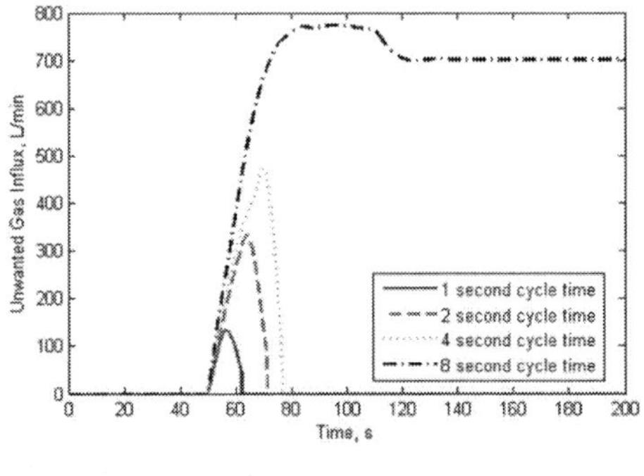

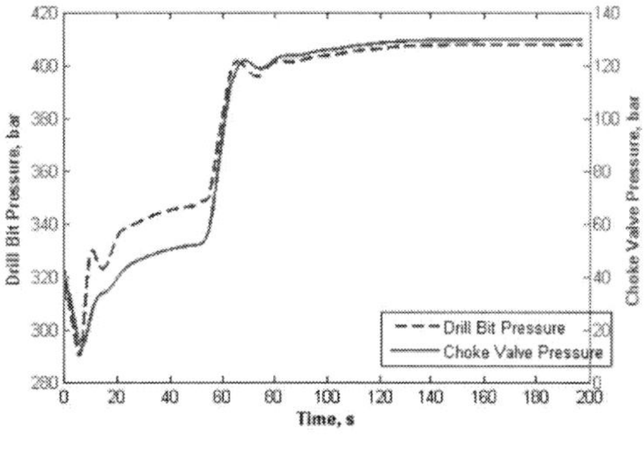

Apr 2, 2014 | Pressure Regulation and Kick Attenuation with Wired Drillpipe in MPD

Thank You for Your Attention

Apr 2, 2014 Pressure Regulation and Kick Attenuation with Wired Drillpipe in MPD

14AIChE®
Spring Meeting, New Orleans, LA

Risks from Unconventional Wells Gasification

Dennis B. Butts
Americas Risk Practice
Environmental Resources Management (ERM)
CityCentre Four
840 West Sam Houston Parkway North, Suite 600, Houston, TX 77024
dennis.butts@erm.com

Prepared for Presentation at
American Institute of Chemical Engineers
2014 Spring Meeting
Topical 2: Shale Gas and Tight Oil
New Orleans, LA
March 30 – April 2, 2014

UNPUBLISHED

Risks from Unconventional Wells Gasification

Dennis B. Butts
Americas Risk Practice
Environmental Resources Management (ERM)
CityCentre Four
840 West Sam Houston Parkway North, Suite 600, Houston, TX 77024
dennis.butts@erm.com

Keywords: LNG, Risk Management, Process Safety, Unconventional Wells, Shale

Abstract

With the continued differential between diesel and natural gas costs, onshore well operators have been working to convert both drilling rigs and hydraulic fracturing pumps to utilize natural gas as fuel. Associated with this conversion is a change in the risk profile that may not be fully recognized or understood in the field. Risk management philosophies and assumptions need to adjust to the difference in hazard characteristics between the fuels, as well as the variety of delivery methods.

Specific topics will include:

- Motivations for fuel conversion and the differences in scale for rig power, natural gas substitution on diesel hydraulic fracturing pumps, and dedicated natural gas powered hydraulic fracturing pumps. Risk mitigation and field training requirements may not fully accounted for in investment decisions.
- Scale of conversion and location specific availability drives natural gas sourcing, which can then drive personnel risk and well pad layout issues depending on whether the gas is sourced from local wellheads, pipeline, Compressed Natural Gas (CNG), or Liquefied Natural Gas (LNG). Separate Simultaneous Operations (SIMOPS) issues can arise based on the specifics of each sourcing method.
- Associated risks need to be appropriately managed in a manner that involves the personnel directly exposed. And there is a need to ensure that well sites which did not previously present a threat to surrounding activities do not become a threat due to either direct risk from the fuel conversion itself or from escalation risks not previously of concern.

1. Introduction

Over the past few years there has been a significant increase in efforts to substitute natural gas for diesel to provide fuel for drilling and fracturing onshore wells. This is an expansion of the conversion efforts that were already underway to convert transportation fleets away from diesel fuel. While the reasons for this may vary, and do include environmental benefits, the most significant reason is cost savings.

Natural gas has been averaging about $4.50/mmbtu [1] over the past 6 months, which is about $0.59 per diesel gallon equivalent (DGE) [1]. Even with added distribution costs, if pipeline gas is available, the economics are very attractive. Wellhead gas is even more attractive. And if LNG is the only alternative at a well site, at $4.50/mmbtu, the natural gas component of LNG is about $0.34 per gallon [1], with a markup of about $0.15 per gallon for liquefaction and storage [3] plus transport costs of about $0.15 per gallon (up to 100 miles) [4], giving a breakeven price of $0.64 per gallon or $1.10 per DGE. With typical supplier pricing to bulk customers as a discount to diesel typically around $1.50 per gallon, and more substantial discounts for large scale contracts, there are significant financial benefits for both LNG suppliers and drillers when diesel fuel is in the $4 per gallon range.

And as a bonus to the economic justification, drillers can tout emissions reductions of as much as 20 – 30% relative to diesel [5]. Rarely in these discussions is the subject of added costs or effort for process and technical safety brought up.

Substitution of natural gas for diesel has been underway for some time in heavy vehicle transportation fleets. With added competition in engine conversion markets and among storage tank suppliers, conversion costs have been reduced to the point where heavy vehicles can recover costs in less than 2 years [6]. Efforts to convert drill rigs originated in the Marcellus region, due to environmental concerns, but the economics have driven natural gas rigs into most of the major markets in the US, as well as into western Canada. Since hydraulic fracturing pumps use even more diesel than rigs, recent efforts have been in this direction, though technical challenges regarding power delivery have resulted in two types of conversion. One is the substitution of gas for diesel in existing engine drivers to ensure sufficient horsepower. The other is to use high powered turbine pumps, which are very well suited to gas power. Large scale conversion of fuel sources is not unprecedented, as the US heavy truck fleet went from less than 10% diesel powered to nearly 100% diesel powered from 1950 to 1980 [6].

So what does this mean in terms of risk management? Aside from the business and financial risk of converting equipment to a fuel in natural gas that is historically more volatile in pricing than diesel fuel at a time of historically high oil to gas price differentials, which is a subject for another paper, there are practical technical safety considerations that come into play as drilling sites convert from diesel to natural gas. These can result in increases to the risk profile of a well pad that is not generally accounted for, as well as changes in the risk profile within the drilling operation itself.

This paper will discuss some of the issues and options regarding conversion and highlight specific risk issues associated with each.

2. Sources of Natural Gas

2.1 Typical Volume Requirements

A heavy truck may use 20,000 gallons of diesel per year [4], with significant potential savings being the result of a conversion to natural gas fuel. Conventional fluid drill rigs use an average of 425,000 gallons of diesel per year [5], creating a fuel cost savings equivalent to converting a fleet of over 20 heavy trucks, but at an investment cost that is much lower. But even these volumes pale in comparison to the amounts of diesel used in hydraulic fracturing. Fracking crews can use 5,000 to 10,000 gallons of diesel per day [7], and it is common for the crews to be in operation for as many as 300 days per year, accounting for mobilization and demobilization time. That means 1.5 to 3 million gallons of diesel per year can easily be used by a single crew, often going well to well for the same operator. An unfortunate outcome of this is that conversions become a foregone conclusion, without due consideration to the resulting process safety risks.

2.2 Options for Sourcing and Delivery

Natural gas for a well site can be sourced as gas directly from the wellhead or gathering system, or it can be piped in from a commercial distribution line, when those are available. In more remote locations, distribution lines may not be available, and if the gas is too wet, wellhead or gathering gas may also be unsuitable for use. In those situations the options are to utilize either CNG or LNG. CNG would require considerably more transportation than LNG due to the lesser amount of gas that can be moved and comes with higher costs for storage and handling equipment. Therefore, CNG has limited utility for well sites, though its lower fuel cost makes it useful in some narrow applications. The use of LNG can be cost effective when other options are not available, as it is more available than in the past and although containing less energy than diesel per gallon, an onsite tank containing 15,000 gallons of LNG is still equivalent to over 8600 gallons of diesel [2] and is about 2.4 times the energy density of CNG [9].

2.3 Issues Associated with Each Option

Use of wellhead gas or gas from gathering lines requires either very dry gas or extensive treatment, as most engines converted to utilize natural gas are designed for gas approximating domestic gas. This is in contrast to offshore facilities where engines are often optimized to take advantage of fuel gas diverted from separation trains. Pipeline gas can be the best option, for those sites with ready access to a distribution line, and will provide sufficient flow and pressure for most natural gas drilling rigs. However, pipeline gas may have to have compression added for certain types of fracking pumps, and available flow may not be sufficient for some fracking operations. CNG can be useful in areas that do not have dry gas or ready pipeline access and may not have easy access to LNG. In such areas, CNG facilities can be set up to provide the product to users in the surrounding region much more readily than LNG facilities. To utilize CNG does require more transportation of the gas and more onsite storage than using LNG. With LNG being the last supply option, but becoming very common in the field, this is an option that carries significant process and technical safety risks. LNG is a material than can bring many benefits, but needs to be treated with considerable respect, often more than is witnessed at well sites.

3. Associated Risks

Associated process and technical safety risks arise in several stages of the conversion from diesel to natural gas. In most cases there is an increase in those risk levels relative to diesel fuel, though this is not always fully contemplated by decision makers or well communicated to the field level staff. Whereas diesel fuel is classified as a combustible liquid, NFPA Class II [8], is difficult to ignite if spilled and primarily a localized threat, natural gas is highly flammable vapor and depending on delivery method could have the potential to disperse considerable distances if released.

In use in equipment, natural gas can create the need for hazardous area class or hot work permit zones that were not required around diesel equipment. This issue has been missed on well sites and often is not included in discussions about conversion costs until after being identified in PHA studies or other risk assessments. The same consideration extends to gas monitoring and shutdown systems. System risks are manageable, but need to be given due consideration and treated with the proper respect. Some expansion of this thought is provided below.

3.1 Source of Gas

Wellhead gas primarily introduces risk through the separation and dehydration process. Even relatively dry gas needs to be treated prior to use in most of the engine drivers used for onshore production. To utilize it during drilling usually means that at least one well has already been put into operation on a multi-well pad. The risks associated with production while drilling and completing adjacent wells have to be assessed. Controls that may need to be in place during certain phases of drilling could interfere with the supply of gas. An alternative that may be available within a field of well sites under one operator is to use gas from the gathering system. While treatment would also be required for that gas, the potential SIMOPS issues of a producing well on a pad with drilling activity. In either case the gas being treated will be a relatively high pressure and is likely to result in some amount of condensate or natural gas liquids (NGL) production and handling. This can increase the fire and explosion risks above those normally found during drilling activity.

Pipeline gas is only an option when a distribution line is nearby. But if available, this can provide a supply of natural gas that is already treated and can be adjusted to the necessary pressure at the takeoff location from the distribution network. In analysis of fire and explosion risks, pipeline gas supply most often has the smallest consequence zones and the least amount of equipment susceptible to failures. This is even the case if high horsepower fracking pumps are in use that requires some degree of compression of the natural gas to reach necessary operating pressures.

CNG introduces risk due to its high pressure, typically near or above 3000 psig [9]. This can both drive the spread of a gas cloud in the event of a release as well as provide the means through depressurization to cool the gas to a low enough temperature that it will disperse for some distance along the ground, further lengthening dispersion distances. Storage under pressure also introduces a finite risk of boiling liquid expanding vapor explosion (BLEVE) that could spread debris.

LNG usage at well sites is capable of imposing a larger potential impact zone than a well blowout scenario. And this is excluding a catastrophic failure of the storage medium, which is low likelihood event due to the dual tank construction of trailers that handle LNG. When released, LNG will initially disperse in a manner very similar to liquefied petroleum gas (LPG), but as the edges of the dispersing cloud warm, flammable vapors will loft upward. Assumptions regarding layout and spacing that are based on LPG can address the flammable breadth of a possible LNG cloud, an example of which is shown in Figure 1 from PHAST [10], going out about 850 feet (260 meters) to the 50% lower flammable limit (LFL). Maximum distance to a 1 psi overpressure, good for window breakage, trailer damage, and upsetting neighbors, would extend another 300+ feet (100 meters) beyond the 50% LFL distance. In general, the height of the 50% LFL section of the cloud will remain under 15 feet (5 meters), but this also needs to be confirmed relative to the drill rig flare height. And the failure of a LNG loading hose would impose an even larger impact zone. Though many operators end up using gas detection, this is only effective if appropriate emergency response procedures are in place.

Figure 1: 50% LFL Dispersion Distance for 1-inch LNG Leak [10]

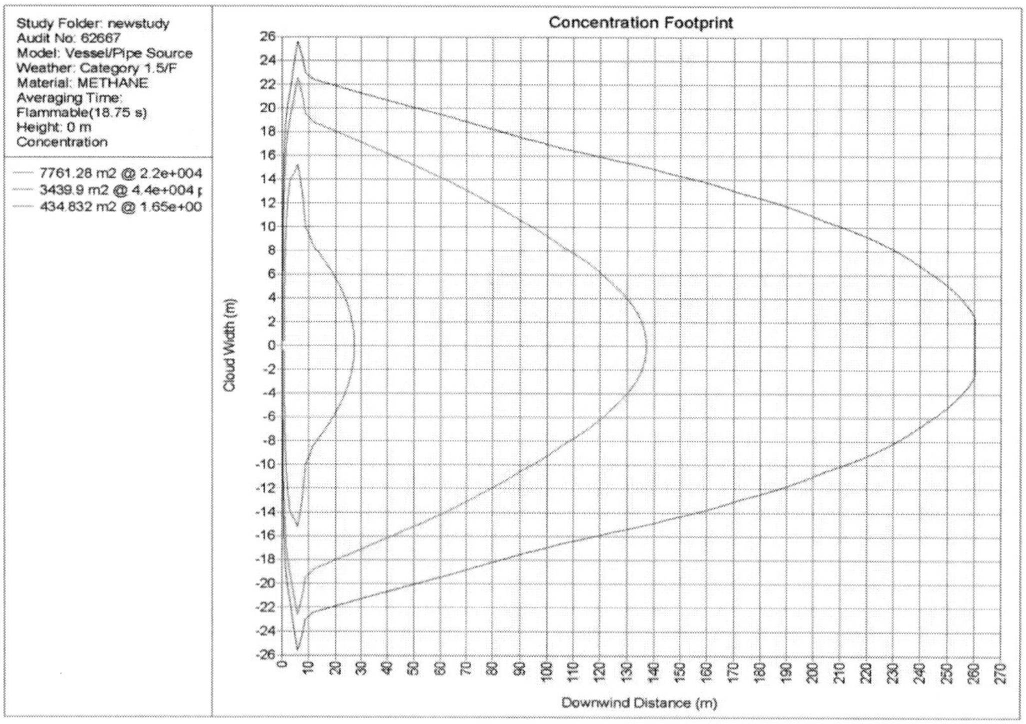

For LNG most releases behave as impinged releases, due to the low pressure environment. For CNG, the best case is for a free field release, which provide dispersion distances of less than half that of LNG. However, horizontal releases of CNG are capable of dispersion distances beyond those of LNG, in particular if they impact other equipment and become partially impinged, as shown for dispersion to 50% LFL for a 1-inch CNG release in **Figure 2**. An interesting phenomenon is how the LNG dispersion slumps outward from the beginning, due to release at temperatures low enough for the methane to be denser than air, but the CNG is initially released at warmer temperatures, then settles into a much lower temperature, resulting in a slumping spread.

Figure 2: 50% LFL Dispersion Distance for 1-inch CNG Horizontal Leak [10]

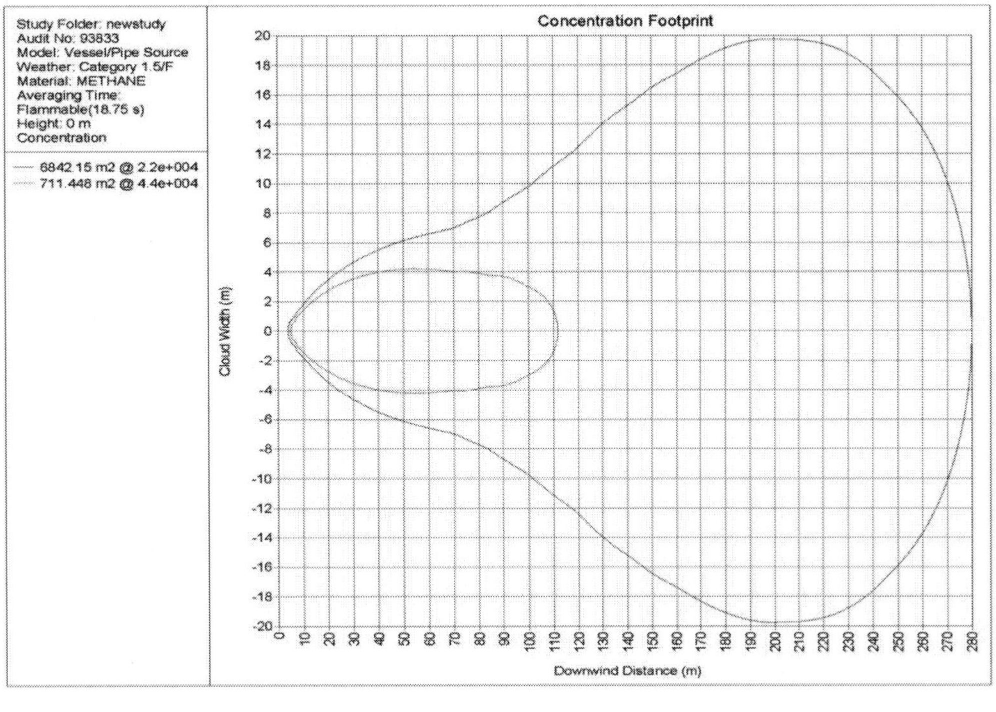

As a reference, an ignited well blowout, traditionally considered a worst case event at a well pad, will typically have an effect distance to 1500 Btu/hr-ft^2 (4.7 kW/m^2) of about 500 feet (150 meters).

3.2 Refueling at Site

When utilizing CNG or LNG as the source of gas, there are additional considerations for storage at the site and for refueling deliveries. Both require delivery by truck, just like diesel fuel, but the release of diesel from an overturned truck is not the same as a release of compressed or liquefied natural gas. This is why there are stringent design requirements on CNG and LNG transport vehicles. However, when assessing risks it is wise to keep in mind that those standards were developed for paved highway transportation, not gravel access roads to well sites. In addition to possible visibility issues, there are the impacts such roads could have on the expected behavior of safety features on the vehicles.

3.3 Compression and Delivery of Gas to Equipment

Further issues arise in situations where either pipeline gas or LNG is used to supply equipment such as hydraulic fracturing pumps that require higher pressure feed gas. In these circumstances, arrangements have to be made to provide compression at the site. A site that is typically already very crowded with equipment. In terms of risk from the equipment it is normally a localized threat, however the extra equipment adds a level of complexity that can complicate operations. The added complexity can add a human factors risk dimension that is rarely assessed in drilling situations.

4. Conclusion

Considerable financial and environmental benefits can be had by converting drilling operations from diesel fuel to natural gas, as long as the current differential in crude oil versus natural gas pricing exists. To achieve those benefits does entail some risk, with this paper having focused on the process and technical safety risks. In some instances, the benefits and risks can pull in opposing directions creating a challenge to find equilibrium between the two.

Depending on the source of natural gas for the drilling and completions operations, due consideration needs to be given to:

- Training needs for field staff that are used to dealing with combustible diesel fuel and not highly flammable natural gas;
- Flammable gas dispersion distances as well as resulting maximum potential zones for 1 psi overpressures, in particular for LNG or CNG applications;
- Review of drill rig flare location relative to natural gas supply point(s) to account for potential height of dispersion;
- Road conditions and their potential impact on transportation of fuels;
- Hazardous area classification zones and resulting equipment alterations, layout changes, or work permit requirements;
- Gas detection needs and the emergency response planning to accompany the detection; and
- Hazard assessment requirements and when they are activated to address changes made to drilling operations or well pad layouts once in service.

As we further explore this question, it is necessary to keep in mind not only the benefits of the ongoing process of converting away from diesel fuel to natural gas at well sites, but also the risk management challenges that can arise.

5. References

[1] US Energy Information Administration, 2014

[2] US Department of Energy, Fuel Energy Spreadsheet, 2014

[3] Quine, Tom, "Fundamentals in LNG", ASGMT White Papers, 2010

[4] Center for Energy Economics, University of Texas at Austin

[5] Oil & Gas Technology, "Gas Powers Fracking Plays", April 10, 2013

[6] Blomerus, Paul and Oulette, Patric, "LNG As a Fuel for Demanding High Horsepower Engine Applications: Technology and Approaches",

[7] Schutt, Bryan, "Diesel to Gas Switching Trend Takes Off in Drilling Sector", SNL Financial, September 3, 2013

[8] NFPA 30, National Fire Protection Association, 2012

[9] Envocare Ltd., "Liquefied Petroleum Gas (LPG), Liquefied Natural Gas (LNG), and Compressed Natural Gas (CNG)", 2007

[10] PHAST 6.7, DNV GL AS

353108 Supercritical CO_2 / Brine / Hydrocarbon Interactions in Shale Rock Fracture: A Geo-Microfluidics Approach

Abstract:

Shale oil and gas is a critical unconventional resource for the energy sector in the US and worldwide. The booming of its production in the recent decade, due to the hydraulic fracturing technique, has led to serious environmental concerns such as huge water demands and toxic additives into fracking water which may pollute groundwater. A proposed solution is to use supercritical CO_2 ($scCO_2$) as a more environment-friendly alternative fracking fluid to replace water for enhanced oil recovery (EOR) in oil shale. Many challenges lie in the way of this scheme such as to understand multiphase flow of $scCO_2$ at injection sites, its interactions with brines, hydrocarbons, and surrounding geologic matrix. We present here a microfluidic approach to investigate the micro-scale multiphase fluid flow and reactive transport within low-permeability shale rock. Unlike the conventional microfluidic techniques, the geo-microfluidic chip is directly fabricated from a thin-section of shale rock, hence provides the genuine surface properties. Simple fracture-like patterns are created on the shale rock surface by machining and femto-second laser etching to simulate the natural structures. Flow experiments in the geo-microfluidic chip are conducted within a high-pressure, high-temperature apparatus to simulate the in situ reservoir conditions. The interactions between $scCO_2$/brine/hydrocarbon will be quantified and the hydrocarbon extraction efficiency of using $scCO_2$ and using fracking water will be compared. The displacement patterns, the dissolution rate of $scCO_2$ in brine/hydrocarbon, and the surface wettability of $scCO_2$ and shale rock will also be investigated. These micro-scale experiments results will help to reveal the underlying mechanisms of multi-phase fluid flow under in situ temperatures and pressures.

352468 Model Shale Gas Properties from a Statistical Mechanical Perturbed Hard-Sphere Chain Cubic Equation of State

Abstract:

Shale gas is an increasingly important energy source in North America. Hence, an accurate description of the physical, transport and thermodynamic properties of shale gas mixtures are a key prerequisite for process design. A particularly advantageous feature is that multiple properties can be obtained from a single equation of state (EOS) for a model system. Traditional EOS's are cubic in molar volume and are derivatives of the basic van der Waals (1873) EOS. Redlich and Kwong (1949), Soave (1973) and Peng and Robinson (1976) are attempts to improve the accuracy of the basic vdW EOS; however, as is shown early in this talk, these equations of state overcorrect the attractive term to compensate for deficiencies in the repulsive term. On the other hand these EOS's have the advantages of a computationally simple form and *three well-defined molar volume roots*. Nonetheless, while these EOS's often give a reasonably accurate representation of the phase composition of simple binary and ternary alkane mixtures (model shale gas systems); poor results are obtained for other properties including liquid density, mixture second virial coefficient and diffusion coefficient.

On the other hand, a class of accurate molecularly-based but complex and computationally intense equations of state: the statistical associating fluid theory (SAFT) equations (Huang and Radosz, 1990; Chapman et. al., 1990), which account for individual molecular level effects (i.e., hard-sphere repulsion, London dispersion, covalent bond chain formation, multipolar forces and hydrogen bonding) have successfully emerged in the past few decades to account for multiple effects in pure fluids and mixtures including associating and/or polydisperse components. These EOS's capture the major molecular level effects; however, convergence difficulties and the existence of several molar volume roots complicate implementation of these EOS routines.

Consequently, considering the above difficulties, we have developed an approach which incorporates the rich molecular description of the SAFT equations into a cubic in molar volume form. Specifically, molecular simulation data for hard-sphere compressibility factor is nearly exactly fit to a simple empirical function in packing fraction, which ensures an overall cubic in molar volume equation of state and from which two universal parameters (applicable to all fluids and fluid mixtures) are obtained. Furthermore, London dispersion attractive forces are described by an extremely simple polynomial refit of data generated for Chen and Kreglewski's (1977) *thirty-six term* double sum in reciprocal absolute temperature and packing fraction. Finally, chain formation (covalent bonding) is described by a modified form of Wertheim's (1984) thermodynamic first-order perturbation theory (TPT1) for segment covalent bonding, in which a diimer reference radial distribution function is employed to obtain more accurate results. Conventional forms of the hard-dimer radial distribution function (Chiew, 1991; Yethiraj and Hall, 1992; Ghonasgi and Chapman, 1994; Sadus, 1999) were refit to a novel empirical functional form which results in an equation of state which is overall cubic in molar volume.

The resulting theoretically-based EOS was applied to predict several model shale gas properties including compressibility factor versus packing fraction isotherms, the vapor-liquid coexistence curves, the second-virial coefficient versus temperature and the self-diffusion coefficient versus pressure isotherms for the constituent pure fluids; and vapor-liquid equilibria and binary diffusion coefficients versus pressure for binary systems. The observed performance of the molecularly-based *cubic* EOS presented in this paper is generally better than the performance of conventional cubic equations of state and about equal to the performance of the more complex non-cubic SAFT equations of state for the same fluids but with much less computational effort. Ongoing work involves development of *cubic* EOS terms which account for multipolar and hydrogen bonding effect to include water and other species in the shale gas mixtures.

REFERENCES

Chapman, W. G., K. E. Gubbins, G. Jackson and M. Radosz, *New Reference Equation of State for Associating Liquids*, Ind. Eng. Chem. Res. **29**, 1709-1721 (1990).

Chen, S. S. and A. Kreglewski, *Applications of the Augment van der Waals Theory of Fluids: I. Pure Fluids*, Berichte der Bunsengesellschaft fur Physikalische Chemie, **81**, 1048-1052 (1977).

Chiew, Y. C., *Percus-Yevick Integral Equation Theory for Athermal Hard-Sphere Chains. II. Average Intermolecular Correlation Functions*. Molec. Phys. **73**, 359-373 (1991).

Ghonasgi, D. and W. G. Chapman, *A New Equation of State for Hard-Chain Molecules*, J. Chem. Phys. **100**, 6633-6639 (1994).

Huang, S. H. and M. Radosz, *Equation of State for Small, Large, Polydisperse and Associating Molecules*, Ind. Eng. Chem. Res., **29**, 2284-2294 (1990).

Peng, D.-Y. and D. B. Robinson, *A New Two-Constant Equation of State*, Ind. Eng. Chem., **15**, 59-64 (1976).

Redlich, O. and J. N. S. Kwong, *On the Thermodynamics of Solutions. V. An Equation of State. Fugacities of Gaseous Solutions*, Chem. Rev. **44**, 233-244 (1949).

Sadus, R. J., *A Simple Equation of State for Hard-Sphere Chains*, AIChE J., **45**, 2454-2457 (1999).

Soave, G. *Equilibrium Constants from a Modified Redlich-Kwong Equation of State*. Chem. Eng. Sci., **27**, 1197-1203 (1972).

van der Waals, J. D., *Over de Continuiteit van den Gas- en Vloeistoftoestand (On the Continuity of the Gas and Liquid State)*. Ph.D. thesis, Leiden, The Netherlands, (1873).

Wertheim, M. S., *Fluids with Highly Directional Attractive Forces I. Statistical Thermodynamics*, J. Stat. Phys., **35**,

19-34 (1984).

Yethiraj, A. and C. K. Hall, *Monte Carlo Simulations and Integral Equation Theory for Microscopic Correlations in Polymeric Fluids*, J. Chem. Phys., **96**, 797-809 (1992).

AUTHOR INDEX

AbdulWahhab, Hamad....71
Adewole, Jimoh....44
Al Khaldi, Saad....125
Al-Ahmed, Amir....44
Al-Harthi, MA....71
Alshaaer, Mazen....71
AlShoaibi, Ahmed....67
Anneken, David....66
Bai, Chuansheng....37
Bearrow, Mike....16
Beebe, Dustin....1
Butts, Dennis....115
Cai, Zhiyong....21
Cao, Xijia....68
Carey, J. W....124
Carrillo, Alvaro....101
Chikamatsu, Nobuyasu....31
Clark, Darren....76
Coppella, Steven....38
Delgass, W.N....37
Donahue, Mike....40
Drayton, Roy A....102
Elsharkawy, Adel....22
Emonds, Mark V.M....77
Enright, Tom....90, 100
Faqir, Naim....71
Fedor, Patrick....24
Fellner, Joseph P....66
Fu, Jinxia....72
Gow, Arthur S....125
Graeme, Ellis....73
Haider, Mohammad Faruq....52
He, Peng....124
Hedengren, John D....103
Honda, Kazunori....31
Hua, Naomi....40
Johnson, Darin....24
Kannan, Pravin....67
Kelly, Robert....125
King, Alan....74
Krueger, Artur W....36
Laine, Patrick....39
Liu, Guangxin....68
Malley....125
Memmott, James....103
Miller, Jeffrey T....37
Misra, Rahul....23
Missana, Terry....100
Mohanty, Sanat....23
Morita, Mitsuo....31
Nolan, Tom....1
O, Anna....125
Obayashi, Shuji....31
Okita, Atsushi....31
Oyama, Koji....31
Pazmino, Jorge....37
Perez, Hector....103
Pixton, David....103
Porter, Mark L....124
Ribeiro, Fabio H....37
Rice, James W....72
Sandhu, Sarwan S....66
Sandström, Emma....72
Scott, Ethenia....20, 65
Shawabkeh, Reyad....71
Shiau, Lie-Ding....70
Shishavan, Reza Asgharzadeh....103
Smith, Robert....73
Sultan, Abdullah S....44
Sun, Jinsheng....68
Suuberg, Eric M....72
Takahashi, Jumpei....31
Teng, Hou-Guo....70
Trimble, Tyler....125
Viswanathan, Hari....124
Wang, Wenping....68
Yan, Qiangu....21
Zaidi, S. M. Javid....44
Zhang, Jilei....21